Introduction to AMC 8

ESSENTIAL ACADEMY

June 2023

AMC8을 지도하면서 맹목적으로 문제만 푸는 학생들을 많이 만나왔습니다. AMC8 시험은 고득점을 위해서 공부하는 것이 아닌 AMC10/12로 가기 위한 가교 역할을 하는 것입니다. 문제만 풀면서 시험 정리를 하는 것이 아니라 조합, 기하, 정수, 대수로 나눠서 공부를 한다면 나도 모르게 지식이 늘어나는 기쁨을 누릴 수 있을 것입니다.

이 책은 혼자 독학으로도 충분히 공부할 수 있도록 설계되었습니다. 기출문제는 인터넷 어디서든 구할 수 있지만 경시대회에서 출제되는 다양한 문제들을 함께 풀어 보아야 진정한 실력을 쌓을 수 있기에 저는 이 책에서 기존 기출문제뿐 아니라 다양한 경시대회 문제들을 함께 수록하였습니다.

Combinatorics, Algebra, Number Theory, Geometry 순으로 책을 구성했습니다.
수학을 공부하면서 즐거움을 느끼길 바랍니다.

2023년 6월 17일
에센셜 아카데미 원장
이종욱

Throughout my time teaching the AMC8, I've encountered many students who blindly focus on solving problems. The AMC8 exam should not merely be a target for high scores but serve as a stepping-stone toward the AMC10/12. Instead of solely focusing on solving problems for exam preparation, if one categorizes and studies the subjects of Combinatorics, Geometry, Integers, and Algebra, one can experience the joy of unknowingly broadening their knowledge.

This book is designed for independent study, allowing you to learn adequately on your own. While past test questions can be found anywhere on the internet, true skill can only be built by tackling a variety of problems that are presented in competitions. Thus, in this book, we've included not only the original past questions but also various competition problems.

The book is organized in the order of Combinatorics, Algebra, Number Theory, and Geometry. I hope you find joy in studying mathematics.
I hope you find joy in studying mathematics.

June 17, 2023
Director of Essential Academy,
Jongwook Lee

Contents

I. Combinatorics — 8

1 Counting — 9
- 1.1 Counting Basics — 9
- 1.2 Venn Diagrams — 10
- 1.3 Three-Set Venn Diagrams — 10
- 1.4 Gauss Formula — 11
- 1.5 Regrouping the Numbers in a Sum and Product — 11
- 1.6 Practice Problems — 11

2 Permutations — 12
- 2.1 Permutation Definition — 12
- 2.2 Factorials — 12
- 2.3 Permutations Fundamentals — 13
- 2.4 Digit Permutations — 14
- 2.5 Circular Arrangements — 15
- 2.6 Practice Problems — 15

3 Combinations — 17
- 3.1 Combination Definition — 17
- 3.2 Binomial Identity — 18
- 3.3 Tricky Question — 18
- 3.4 Word Arrangement Fundamentals — 19
- 3.5 Word Arrangements with Constraints — 19
- 3.6 Practice Problems — 20

4 Probability — 22
- 4.1 Probability Definition — 22
- 4.2 Probability of Independent Events and Dependent Events — 22
- 4.3 Complementary Counting and Probability — 24
- 4.4 Practice Problems — 24

5 Casework Problems — 26
- 5.1 Practice Problems — 26
- 5.2 Harder Question — 26

6 Principles of Inclusion and Exclusion — 28
- 6.1 PIE Strategy — 28
- 6.2 PIE for 2 Events — 28
- 6.3 PIE for 3 Events — 30
- 6.4 PIE for n Number of Events — 31
- 6.5 Practice Questions — 31

7 Stars and Bars — 34
- 7.1 Stars and Bars Fundamentals — 34
- 7.2 Stars and Bars With Constraints — 35
- 7.3 Practice Questions — 36

8 Geometric Counting — 38
- 8.1 Geometric Counting Fundamentals — 38
- 8.2 Number of Squares in Grid — 38
- 8.3 Number of Rectangles in Grid — 39
- 8.4 Path Counting — 39
- 8.5 Practice Problems — 41

9 Recursion — 43
- 9.1 Recursion Fundamentals — 43
- 9.2 Recursion with Constraints — 44
- 9.3 Probability Recursion — 45
- 9.4 Practice Problems — 45

II. Algebra — 48

10 Ratio and Percentage — 49
- 10.1 Ratio Fundamentals — 49
- 10.2 Rate and Work — 49
- 10.3 Practice Problems — 51

11 Linear Function and Quadratic Function — 54
- 11.1 Linear Function and Quadratic Function — 54
- 11.2 Word Problems — 55
- 11.3 Practice Problems — 56

12 Speed, Distance, and Time — 60
- 12.1 Speed, Distance and Time — 60
- 12.2 Harmonic Mean — 61
- 12.3 Practice Problems — 61

13 Sequence and Series — 64
- 13.1 Arithmetic Sequence — 64
- 13.2 Geometric Sequence — 66
- 13.3 Practice Problems — 66

14 Mean, Median, and Mode — 68
- 14.1 Mean, Median, Mode — 68
- 14.2 Mean, Median Mode Conditions Example — 68
- 14.3 Practice Problems — 70

15 Telescoping — 75
- 15.1 Telescoping — 75
- 15.2 Telescoping Sum and Product — 75
- 15.3 Practice Problems — 76

III. Number Theory — 77

16 Primes and Divisibility — 78
- 16.1 Primes — 78
- 16.2 Divisibility Rules — 78
- 16.3 Primes Factorization — 79
- 16.4 Practice Problems — 82

17 Factors ... 85
- 17.1 Number of Factors ... 85
- 17.2 Sum of Factors ... 85
- 17.3 Product of Factors ... 85
- 17.4 Practice Problems ... 86

18 GCF and LCM ... 88
- 18.1 GCF and LCM Fundamentals ... 88
- 18.2 GCF and LCM Product ... 88
- 18.3 GCF and LCM Properties ... 88
- 18.4 Euclidean Algorithm ... 89
- 18.5 Practice Problems ... 90

19 Modular Arithmetic ... 92
- 19.1 Modular Definition ... 92
- 19.2 Product Rule ... 92
- 19.3 Exponent Rule ... 92
- 19.4 Multiple Modular Congruence ... 93
- 19.5 Digit Cycle ... 93
- 19.6 Practice Problems ... 93

IV. Geometry ... 95

20 Angle Chasing ... 96
- 20.1 Angle Chasing Tricks ... 96
- 20.2 Inscribed Angle ... 98
- 20.3 Polygons ... 98
- 20.4 Advanced Circle Angle Chasing Theorems ... 99
- 20.5 Practice Problems ... 102

21 Triangle ... 108
- 21.1 Area of Triangle ... 108
- 21.2 Special Triangle ... 110
- 21.3 Pythagorean Theorem ... 111
- 21.4 Triangle Properties ... 113
- 21.5 Angle Bisector Theorem ... 114
- 21.6 Practice Problems ... 114

22 Quadrilateral ... 120
- 22.1 Square ... 120
- 22.2 Rectangle ... 120
- 22.3 Rhombus ... 120
- 22.4 Parallelogram ... 121
- 22.5 Trapezoid ... 121
- 22.6 Practice Problems ... 121

23 Circles ... 123
- 23.1 Circle Theorem ... 123
- 23.2 Circular Area ... 123
- 23.3 Length Inside Circles ... 124
- 23.4 Practice Problems ... 125

24 Similarity and Congruence **131**
   24.1 Congruent Triangle 131
   24.2 Similar Triangle 131
   24.3 Practice Problems 132

25 Area of Polygon **135**
   25.1 Area of Polygon 135
   25.2 Practice Problems 136

26 3D Geometry **141**
   26.1 Cube 141
   26.2 Prism 141
   26.3 Pyramid 142
   26.4 Cylinder 143
   26.5 Cone 143
   26.6 Sphere 143
   26.7 Practice Problems 144

Combinatorics

Essential Academy (June 2023) — Introduction to AMC 8

§1 Counting

§1.1 Counting Basics

Example 1.1
Alan takes a pack of index cards and numbers them starting with 20 and ending with 80. How many cards does she number?

Solution. Numbering cards from $20 \sim 80$ is the same as numbering cards from 1 to 61 (subtracting 19 from each number, we get 1 through 61 instead of 20 through 80), so there are 61 cards.

Question. How many positive integers are between a and b exclusive, where a is less than b?

Between integers a and b, there are $(b-a)-1$ integers. Exclusive means that a and b are excluded.

Question. How many numbers are there from a to b inclusive ($a < b$)?

From a to b inclusive there are $(b-a)+1$ integers. Inclusive means that a and b are included.

Example 1.2
Example: Messi owns a rectangular plot of land that is 60 yards long and 30 yards wide.

1. If he places a fence post at each corner and posts are placed every three yards, how many fence posts will there be on each side?

2. If he places a fence post at each corner and posts are placed every three yards, how many fence posts will he need to enclose his land?

Solution. At first glance, the answers to these problems seem to contradict each other. For the first problem, instead of considering the posts, consider the gaps between the posts. To go 60 yards requires twenty gaps. The first post is placed, and then a post is placed every 3 yards for a total of 21 posts. 30 yards requires 10 gaps, or 11 posts.

If two of the sides have 11 posts, and the other two sides have 21, it would appear that there are 64 posts, but we counted all four corners twice. We must subtract four to get a total of 60 posts.

When calculating the total, I prefer to think of it this way:

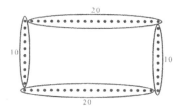

Law of sum and product:

- If events are happened at the same time, then it need to be multiplying.

§1.2 Venn Diagrams

A **venn diagram** can be used to organize counting problems where some items are included in multiple groups and others are excluded.

> **Example 1.3**
> In math classroom: 19 students have an iPhone, 15 students have a galaxy folds, 7 students have a both iPhone and galaxy folds, and 6 students don't have any mobile at all. How many students are in the classroom?

In math classroom: 19 students have an iPhone, 15 students have a galaxy folds, 7 students have a both iPhone and galaxy folds, and 6 students don't have any mobile at all. How many students are in the classroom?

When you fill-in a Venn diagram to solve a problem, work from the inside out. First, we fill-in a 7 where the two groups overlap. Then we can put 12 students who only have iPhone and 8 students who only have Galaxy fold. Six students have no mobiles for a total of $7 + 12 + 8 + 6 = 33$ students.

§1.3 Three-Set Venn Diagrams

> **Example 1.4**
> Every student who applied for admission to a veterinary school has at least one pet: 30 have a cat, 28 have a dog and 26 have fish. If 13 students have fish and a cat, 15 students have fish and a dog, 11 students have both a cat and a dog, and 4 students have a cat, a dog and fish. How many students applied to veterinary school? Begin at the center of the diagram below and work your way out to get:
>
>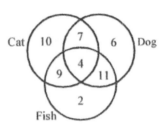

Using a Venn diagram is easy to follow but you use basic counting technique. Use inclusion – exclusion principal for solving. It will be in Chapter 7.

30 have cats + 28 have a dog + 26 have fish = 84 owners. Students who own two (or more) different animals were counted multiple times, so we subtract them: $84 - 13 - 15 - 11 = 45$. But we have to add

students who have all three pets three times. Finally add 4 students to get 49.

§1.4 Gauss Formula

The gauss formula states that the sum of all integer numbers starting at 1 and finishing with n can be found by the following formula

$$1 + 2 + 3 + \cdots + n = \frac{n(n+1)}{2}$$

Sum of Even Numbers:
$$2 + 4 + \cdots + 2n = n(n+1)$$

Sum of Odd Numbers:
$$1 + 3 + \cdots + (2n-1) = n^2$$

§1.5 Regrouping the Numbers in a Sum and Product

Regrouping Sum

Commutative Property: $a + b = b + a$

Associative Property: $(a+b) + c = a + (b+c) = a + b + c$

Regrouping Product

Commutative Property: $a \times b = b \times a$

Associative Property: $(a \times b) \times c = a \times (b \times c) = a \times b \times c$

§1.6 Practice Problems

Problem 1.5
Find the value of $200 - 198 + 196 - 194 + \cdots + 4 - 2$.

(A) 400   **(B)** 100   **(C)** 120   **(D)** 200   **(E)** 50

Problem 1.6
Find the value of $-1 + 2 - 3 + 4 - 5 + 6 - 7 + \cdots - 499 + 500$.

(A) -200   **(B)** 200   **(C)** -250   **(D)** 250   **(E)** 500

§2 Permutations

§2.1 Permutation Definition

A permutation is a possible arrangement of objects in a set where the order of objects matter.

> **Example 2.1**
> How many 3 digit passwords are possible where each digit is a number from 0 to 9?

Step by Step:

First: How many 1 digit passwords are there?

Second: How to follow the number of 2 digit passwords?

Finally: How many 3 digit passwords are there?

§2.2 Factorials

Definition 2.2. A factorial is the product of all positive integers less than or equal to a given positive integer. In other words,
$$n! = n \times (n-1) \times (n-2) \times \cdots \times 1$$

Point: Remarkable factorials in math competition

- $0! = 1$
- $1! = 1$
- $2! = 2$
- $3! = 6$
- $4! = 24$
- $5! = 120$
- $6! = 720$

Definition 2.3. A **permutation** is a possible arrangement of objects in a set where the order of objects matter.

> **Example 2.4**
> Alan has 1 wrapping paper in each of 4 different colors: red, orange, green, or blue. He needs to wrap all of them around a box in any order. How many ways are there for him to do this?

Solution. Please notice that, we can't use the same wrapping paper twice so it's slightly more complicated. Let's begin this problem by approaching each layer at a time.

First, we can choose any of the 4 colors so there are just 4 ways for this to happen.

Question. How many choices for the 2nd layer?

We had 4 choices at the first time, so only 3 of the colors left since we already used one. Therefore, for the 2nd layer we have 3 choices for the color.

For the 3rd layer, we have already used 2 of the wrapping papers, so there are only 2 choices left. For the final layer, we have already used 3 of the papers, so we only have 1 choice for the final paper.

So we can say $4 \times 3 \times 2 \times 1 = 24$, expressed as $4!$.

Example 2.5

Find the number of ways to arrange 4 different books on a bookshelf.

There are 4 different AMC8 Videos. Every student in the class watches the videos in a different order. What is the Maximum number of students in the class?

Factorial
$$n! = n \times (n-1) \times (n-2) \times \cdots \times 1$$

§2.3 Permutations Fundamentals

Example 2.6

Sam has 5 different stamps and is making a Christmas card where he puts 3 different stamps in a line. How many different Christmas card designs can he make?

Definition 2.7. (Permutations Formula) The number of ways to order k objects out of n total objects is
$$_nP_r = n \times (n-1) \times \cdots \times (n-(r-1)) = \frac{n!}{(n-r)!}$$

Example 2.8

Compute

1. $_7P_3$
2. $_5P_2$

Solution.
$$_7P_3 = \frac{7!}{(7-3)!} = 7 \times (7-1) \times \cdots \times (7-(3-1)) = 7 \times 6 \times 5 = 210$$

Solution.
$$_5P_2 = \frac{5!}{(5-2)!} = 7 \times (5-1) \times \cdots \times (5-(2-1)) = 5 \times 4 = 20$$

> **Example 2.9**
> Find the number of ways of selecting a president, vice president, and a secretary from a group of 7 people.

Solution. Let's consider the number of ways for each position separately.

There are 7 choices for the president. Now, because there are only 6 people left, there are 6 choices for the vice president. Then, because 2 of the people are already chosen for the president and vice president, there are 5 choices for the secretary.

In total, the number of ways is $7 \times 6 \times 5 = 210$

§2.4 Digit Permutations

> **Example 2.10**
> How many 4-digit numbers have all digits distinct?

Solution. Let's consider the number of ways to form this number digit by digit.

For the first digit, how many choices do we have? Remember, the first digit of a number cannot be 0 so there are 9 choices (any number from 1 to 9).

How do we handle the condition that the numbers must have distinct digits?

However, unlike the first digit, the 2nd digit can be 0. The 2nd digit can be any number from 0 to 9 except the digit chosen as the first digit, so there are $10 - 1 = 9$ choices.

For the 3rd digit, again, it can be anything from 0 to 9 except the 2 digits already chosen, so there are $10 - 2 = 8$ choices. For the final digit, there are $10 - 3 = 7$ choices.

So, the total number of possible numbers is $9 \times 9 \times 8 \times 7 = 4536$

> **Example 2.11**
> How many 4-digit numbers exist such that the first digit is odd and the other 3 digits are even and all digits distinct?

Solution. The first digit has 5 choices (1, 3, 5, 7, or 9). The 2nd digit also has 5 choices (0, 2, 4, 6, 8).

Do we have to subtract any of the choices of the 2nd digit to make sure the numbers are distinct?

Keep in mind that although the digits have to be distinct, the 2nd digit is even and the first digit is odd, so there is no overlap. The 3rd digit can be any even number except the one chosen for the 2nd digit, so it has 4 choices.

The 4th digit can be any even number except those chosen for the 2nd and 3rd digit.

In total, the number of 4 digit numbers with these constraints is $5 \times 5 \times 4 \times 3 = 300$.

Example 2.12

How many 5-digit palindromes are there? A palindrome is a number that reads the same forward and backward.

Example 2.13

How many 6-digit numbers exist such that all the digits are distinct, and it's first 2 digits are 6 or more, the last 2 digits are 5 or less, and the 3rd digit is a nonzero multiple of 7?

---

§2.5 Circular Arrangements

The number of ways of arranging n objects in a circle where rotations of the same arrangement aren't considered distinct is
$$(n-1)!$$
The number of ways of arranging n objects in a circle where rotations and the reflections of the same arrangement aren't considered distinct is
$$\frac{(n-1)!}{2}$$

Example 2.14

How many ways are there to arrange 5 people in a circle if rotations are not counted as distinct orientations?

Solution. When things are arranged in a circle, we can fix the position of one person, and look for arrangements of others.

How many other people are left to arrange?

We can simply apply the formula for 5 people where rotations don't matter. This is just $(5-1)! = 4! = 24$

Example 2.15

How many ways are there to arrange 5 people around a circle such that 2 of the people, Alex and Bob, must sit next to each other. Note that rotations are not counted as distinct orientations.

---

§2.6 Practice Problems

Problem 2.16
For the pick 3 lottery, six balls numbered 1 through 6 are placed in a hopper and randomly selected one at a time without replacement to create a three-digit number. How many different three-digit numbers can be created?

Problem 2.16

Problem 2.17
How many ways can five different books be ordered on a shelf from left to right?

Problem 2.18
The eight members of student council are asked to select a leadership team: president, vice-president, and secretary. How many different leadership teams are possible?

Problem 2.19
You are ranking your elective class choices for the upcoming school year. You must rank your top 5 selections in order of preference. How many ways can you rank 5 of the 7 electives that interest ways can you rank 5 of the 7 electives that interest you?

Problem 2.20
Find the number of arrangements of the letters in the word FOOTBALL.

Problem 2.21
Find the number of arrangements of the letters in the word CAMERA.

Problem 2.22
Find the number of arrangements of the letters in the word LADDER.

Problem 2.23
Find the number of arrangements of the letters in the word BANANA.

Problem 2.24
Seven students line up on stage. If Victor insists on standing next to Chloe, how many ways are there to arrange the students on stage from left to right?

Problem 2.25
How many arrangements of the letters in the word ORDERED include the word RED?

§3 Combinations

§3.1 Combination Definition

> **Example 3.1**
> How many ways can I select 3 pencils from 6 different pencils?

Let's try to approach this similarly to permutations. There are 6 choices for the first pencil, 5 choices for the 2nd pencil, and 4 choices for the 3rd pencil.

So the total number of ways is just $6 \times 5 \times 4 = 120$.

Is this the number of ways of choosing or arranging?

Remember that we are just selecting 3 pencils, so the order doesn't matter. Therefore, we are overcounting many cases. For example, choosing pencil A first and pencil B second is the same as choosing pencil B first and pencil A second.

So how do we account for this?

Well, to answer that question, we must consider how many ways there are to order the pencils. Let's say we have 3 pencils.

How many ways are there to order them?

This is just 3! because we are arranging 3 objects in order. Therefore, we are overcounting by a factor of 3!. So the total number of ways to "choose" 3 pencils from 6 different pencils is

$$\frac{6 \times 5 \times 4}{3!} = \frac{6!}{3!3!}$$

Definition 3.2. (Combinations Formula) The number of ways to choose k objects out of a total of n objects is

$$_nC_k = \frac{n!}{k!(n-k)!}$$

Relation Between Permutation and Combination.

> **Example 3.3**
> You are working on the Mastering AMC 8 Book, and you must choose 7 of the 10 combinatorics chapters to work on. How many ways are there to do this?

> **Example 3.4**
> You are working on the Mastering AMC 8 Book, and you must choose 7 of the 10 combinatorics chapters to work on. However, the first 2 chapters are required to understand the remaining chapters so they must be chosen. How many ways are there to do this?

§3.2 Binomial Identity

Theorem 3.5
(Binomial Identity)
$$_nC_0 + {_nC_1} + {_nC_2} + \cdots + {_nC_n} = 2^n$$

Example 3.6
Alan has an apple, orange, banana, pear, and raspberry. He wants to take a fruit basket to a picnic. How many different types of fruit baskets can he take using the fruits he has?

Solution. We can do this by listing the number of ways of choosing all possible combination of fruits.

Is there any easier way to solve this question?

Notice that for each fruit, we have 2 choices. You can either take it, or leave it. We then have to multiply this for all of the 5 fruits. Therefore, the number of ways is simply
$$_5C_0 + {_5C_1} + {_5C_2} + \cdots + {_5C_5} = 32$$

Remark 3.7. Note that one of the fruit baskets will have none of the fruits (which is mathematically a possible type of fruit basket and the problem statement does not explicitly says that it is not allowed).

Example 3.8
You have 5 different pencils and 3 different erasers and must choose some of them to take to the AMC 8. How many ways are there to do this?

§3.3 Tricky Question

Example 3.9
There are 6 experienced applicants and 7 inexperienced applicants applying for a job. Out of the experienced applicants, 2 managers are selected. Out of all of the other applicants who aren't selected for a manager (both experienced and inexperienced), 4 other employees are selected. How many ways are there to do this?

Solution. To solve this problem, we will find the number of ways of choosing managers and other employees separately.

First, how many ways are there to select 2 managers? Since they must be from the 6 experienced applicants, there are
$$_6C_2 = 15.$$
Next, how many ways are there to select the 5 other employees?

Since 2 of the experienced applicants were already selected, there are 4 experienced applicants and 7 inexperienced applicants left. Therefore, we must select 4 employees out of the 11 remaining people.

We can do this in
$$_{11}C_4 = \frac{11!}{4!7!} = \frac{11 \times 10 \times 9 \times 8}{4 \times 3 \times 2 \times 1} = 330$$
Therefore, in total, how many ways are there to select the 3 managers and 5 other employees?

Since there are 15 ways to select the managers and 330 ways to select the other employees, there will be $15 \times 330 = 4950$ ways to select both the managers and the employees.

Example 3.10

You have 12 different energy bars that you need to give to 3 people: Alice, Betty, and Chase. Alice needs 3 bars, Betty needs 4 bars, and Chase needs 5 bars. How many ways are there to distribute the 12 bars to satisfy their requirements?

Example 3.11

How many ways are there to place 4 balls into a 4×6 grid such that no column or row has more than one ball in it? (Rotations and reflections are considered distinct.)

---

§3.4 Word Arrangement Fundamentals

Example 3.12

How many ways are there to arrange the letters in the word ESSENTIAL?

Example 3.13

How many ways are there to arrange the letters in the word COCACOLA?

Example 3.14

How many ways are there to arrange the letters of the word APPLE?

Theorem 3.15

(Word Rearrangements) The number of ways to order a word is
$$\frac{n!}{a_1! a_2! a_3! \cdots a_k}$$
where n is the length of the word, k is the number of distinct letters, and $a_1, a_2, a_3, \ldots, a_k$ are the number of times each of the duplicate letters that appear in the word. $n = a_1 + a_2 + \ldots + a_k$.

---

§3.5 Word Arrangements with Constraints

Example 3.16
How many ways are there to rearrange the letters in COMPUTER such that the C, O, M, and P stay together (not necessarily in the same order)?

Solution. This problem is slightly different from the standard word rearrangement problems because we now have a constraint.

How should we deal with the condition that the COMP must stay together?

Let's treat COMP as 1 block. Since the letters anyways have to stay together, let call this block a COMP. Now, instead, we can find the number of arrangments of COMPUTER.

How many ways are there to arrange COMPUTER?

Notice that this is just 5! since all the letters (or symbols) are different. Now, are we done, or is there something we forgot to account for?

Remember that letters in the word COMP can appear in any order inside the box, so we have to also account for rearrangements of these letters within the COMP.

For each unique arrangement of COMPUTER, how many ways are there to arrange the letters in the word COMP ?

There are 4! ways to arrange the 4 letters C, O, M, and P. Therefore, in total, there are

$$5! \times 4! = 120 \times 24 = 2880$$

ways to rearrange COMPUTER such that the letters COMP remain together.

Example 3.17
How many ways are there to arrange the letters in LOLLIPOP if the I must be next to both an L and O?

§3.6 Practice Problems

Problem 3.18

How many pentagons can be formed by tracing the lines of the figure below?

Problem 3.19

How many ten-unit paths are there between A and B which do not pass through X?

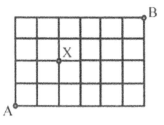

Problem 3.20

Five lines are drawn so that each intersects the other four, but no three lines intersect at the same point. Each of the points of intersection is than connected to every other point of intersection. How many triangles are formed which have three of the points of intersection as vertices?

Problem 3.21

John walks six blocks on a city grid of sidewalks to his favorite deli for lunch: three blocks north and three blocks west. He never uses the exact same path on his return to work. If John always stays on city sidewalks and goes six blocks each way, how many different ways can John walk to the deli and back?

§4 Probability

§4.1 Probability Definition

Definition 4.1.
$$\text{Probability} = \frac{\text{Total Number of Desired Outcomes}}{\text{Total Number of Possible Outcomes}}$$

> **Example 4.2**
> What is the probability of rolling a prime number on a 6 sided dice?

> **Example 4.3**
> What is the probability that the product of 4 dice rolls is a multiple of 648?

§4.2 Probability of Independent Events and Dependent Events

Two events are Independent if the result of the first has no effect on the second. If two events are dependent then the result of the first affects the outcome of the second.

If the probability of an event occurring is $P(A)$, and the probability of second event occurring is $P(B)$, then the probability that both events will occur is:
$$P(A \text{ and } B) = P(A)P(B)$$
If two events are mutually exclusive(meaning both cannot happen), we calculate the sum of the probabilities.
$$P(A \text{ or } B) = P(A) + P(B)$$

> **Example 4.4**
> Messi randomly picks a number from 1 to 10. Ronaldo randomly picks a number from 1 to 25. What is the probability that the product of the numbers they choose is odd?

Solution. For this problem, although it is possible to find the number of successful and total outcomes amongst both picks of numbers, there is an easier way to solve the problem.

To start, what do we know about 2 numbers whose product is odd?

For the product of 2 numbers to be odd, both numbers must be odd because if any of the numbers are even, then the product will also have a factor of 2 in it.

Next, how can find the probability that Messi and Ronaldo pick odd numbers?

We could find the number of successful and total outcomes amongst both picks of numbers as mentioned earlier, but instead, we can simply find the probabilities of each of them picking an odd number.

The probability of Messi picking an odd number is 5/10.

The probability of Ronaldo picking an odd number is 13/25 since there are 13 odd numbers from 1 to 25.

Now, how do we find the overall probability of both picking an odd number?

We must multiply the probabilities.

Therefore, the overall probability is:
$$\frac{1}{2} \times \frac{13}{25} = \frac{13}{50}$$

Example 4.5
Person A rolls a dice with the numbers $1, 1, 2, 6, 15, 30$ and Person B rolls a dice with numbers $1, 2, 4, 18, 28, 44$. What is the probability that the product of the numbers is a multiple of 24?

Dependent Events

Example 4.6
Alex, Betty, Charlie, Derek, Emma, Fiona, and George are racing in a marathon. If they finish in a random order, what is the probability that Charlie is 1st and George is 6th?

Example 4.7
On Saturday, there is a 20% chance of rain. On Sunday, there is a 30% chance of rain if it rained on Saturday, but only a 10% chance of rain if it didn't rain on Saturday. What is the probability that it will rain on both days and neither day?

Example 4.8
(AIME) There is a 40% chance of rain on Saturday and a 30% chance of rain on Sunday. However, it is twice as likely to rain on Sunday if it rains on Saturday than if it does not rain on Saturday. The probability that it rains at least one day this weekend is $\frac{a}{b}$, where a and b are relatively prime positive integers. Find $a + b$.

Mutually Exclusive

Definition 4.9. If events A and B are **mutually exclusive**, then
$$P(A \text{ or } B) = P(A) + P(B)$$
More generally, if $A_1, A_2, \ldots, A_n$ are mutually exclusive events, then
$$P(A_1 \text{ or } A_2 \text{ or } \ldots \text{ or } A_n) = P(A_1) + P(A_2) + \cdots + P(A_n)$$

§4.3 Complementary Counting and Probability

$$P(A \text{ does not occur}) = 1 - P(A \text{ occurs})$$

Example 4.10
The chance of rain for each of the next five days is 25%. What is the probability that it will rain at least once in the next five days? Express your answer to the nearest tenth of a percent.

Example 4.11
I roll 2 fair 6-sided dice. How many ways are there for the sum of the numbers to be 11 or less?

Example 4.12
You have 7 slips of paper numbered 1 to 7. How many ways are there to choose any subset of them so that you have at least 2 odd numbers and 1 even number?

Example 4.13
When a certain unfair die is rolled, an even number is 3 times as likely to appear as an odd number. The die is rolled twice. What is the probability that the sum of the numbers rolled is even? (AMC10)

Example 4.14
I roll 2 fair 6-sided dice. What is the probability that the sum of the numbers is 5 or more?

Example 4.15
How many subsets of the set $\{1, 2, 3, 4, 5, 6, 7, 8, 9, 10\}$ have at least 1 even number and do not contain all of the elements?

§4.4 Practice Problems

Problem 4.16
How many two-digit integers use two different digits?

Problem 4.17
There are seven parking spaces in a row which must be assigned to four co-workers. How many ways can the spaces be assigned if at least two of the assigned spaces must be adjacent?

Example 4.18
How many different ways can six friends stand in line at the movies if Messi and Ronaldo refuse to stand next to each other?

Example 4.19
How many positive integers less than 50 are not divisible by 3 or 5?

Example 4.20
A fair coin is flipped ten times. How many ways are there to flip more heads than tails?

Example 4.21
Five lines are drawn so that each intersects the other four, but no three lines intersect at the same point. Each of the points of intersection is than connected to every other point of intersection. How many triangles are formed which have three of the points of intersection as vertices?

§5 Casework Problems

§5.1 Practice Problems

Example 5.1
Two dice are rolled. What is the number of the ways that the sum of the numbers rolled is 3?

Example 5.2
Two cards are dealt from a deck of four red cards labeled A, B, C, D and four green cards labeled A, B, C, D. Winning pair is two of the same color or two of the same letter. How many ways are there to draw a winning pair?

Example 5.3
Alan has 10 boxes. Each box has a green, orange, red, and blue ball. He randomly chooses one of the boxes and then uniformly at random picks a ball from that box. He repeats this process of randomly choosing any box (allowed to choose same box again) and randomly choosing a ball in that box. What is the probability that the balls are the same color?

§5.2 Harder Question

Example 5.4
There are 3 identical blue boxes, 6 identical green boxes, and 5 distinct items. Each of the items must be placed into either a green or blue box, and each box can contain a maximum of 1 item. Assuming all the boxes are different, how many ways are there to do this?

Solution. There are 5 items to be put in some of the 9 green and blue boxes, so we must choose 5 of these 9 boxes to put the items into.

How can we divide this problem into cases?

We can choose our cases based on the number of green and blue boxes we are selecting. The cases are

1. 3 blue boxes, 2 green boxes
2. 2 blue boxes, 3 green boxes
3. 1 blue box, 4 green boxes
4. 0 blue boxes, 5 green boxes

Notice that this covers all the cases because it covers all possible number of blue boxes. Clearly, 6 green boxes is impossible since we only have to select 5 of them to put items into. Also, make sure to keep in mind that the boxes of the same color are identical.

Case 1: 3 blue boxes, 2 green boxes

For this case, how many ways are there to put 5 items into 3 blue and 2 green boxes?

We can count this be seeing that we simply have to choose 3 of the items to place into the blue boxes, and the other items will be put in green boxes. We don't have to worry about order of the items in the blue boxes (or the green boxes) because the boxes of the same color are identical. We can do this in $_5C_3 = 10$ ways.

Case 2: 2 blue boxes, 3 green boxes Again, we do this by choosing 2 of the items to put into blue boxes, and the remaining 3 items will be put into green boxes. We can do this in $_5C_2$ ways.

Case 3: 1 blue boxes, 4 green boxes Out of the 5 items, we must choose 1 of them for a blue box and the remaining 4 will go in green boxes. We can do this in $_5C_1$ ways.

Case 4: 0 blue boxes, 5 green boxes Because all the items must go in green boxes, there is only 1 way to do this.

In total among all the cases, there are

$$_5C_3 + {_5C_2} + {_5C_1} + 1 = 26$$

§6 Principles of Inclusion and Exclusion

§6.1 PIE Strategy

Overcounting is the process of counting more than what you need and then systematically subtracting the parts which do not belong.

The principle of inclusion and exclusion (PIE) is a counting technique that uses overcounting to compute the number of elements that satisfy at least one of several properties while guaranteeing that elements satisfying more than one property are not counted twice.

§6.2 PIE for 2 Events

> **Example 6.1**
> Find how many numbers from 1 to 100 (inclusive) that are divisible by 2 or 7.

To solve this problem, let's first find how many numbers are divisible by 2 and how many numbers are divisible by 7.

How many numbers from 1 to 100 are divisible by 2?

Every 2nd number is divisible by 2, so all the numbers from 2×1 to 2×50 work. Therefore, from 1 to 100 there are 50 numbers divisible by 2.

How many numbers from 1 to 100 are divisible by 7?

Every 7th number is divisible by 7, so all of the numbers from 7×1 to 7×14 work. We calculate this by seeing that $\frac{100}{7}$ is a little more than 14 so 7×14 is the largest multiple of 7 less than 100. Therefore, there from 1 to 100 are 14 numbers divisible by 7.

Are there any numbers overcounted between the multiples of 2 and the multiples of 7?

For a number to be both a multiple of 2 and a multiple of 7, it must be a multiple of $7 \times 2 = 14$. Therefore, we are overcounting these numbers since we are counting them twice as part of the multiples of 2 and multiples of 7.

How many numbers from 1 to 100 are divisible by 14?

Every 14th number is divisible by 14, so all the numbers from 14×1 to 14×7 work. Therefore, from 1 to 100 there are 7 numbers divisible by 14. Therefore, in total, the number of numbers that are multiple of 2 or 7 are the number of multiples of 2 plus the number of multiples of 7 minus the multiples of 14.

This gives us an answer of $50 + 14 - 7 = 57$.

Union and Intersection: ∩, ∪

Theorem 6.2

(Principle of Inclusion Exclusion for 2 Sets)

$$|A \cup B| = A + B - |A \cap B|$$

Basically, we count the number of possibilities in 2 "things" and subtract the duplicates.

Example 6.3

How many subsets of the set $\{1, 2, 3, 4, 5, 6, 7, 8, 9, 10\}$ have at most 4 even numbers and at most 2 elements that are multiples of 3?

Solution. **How we can apply the Principle of Inclusion and Exclusion to this problem?**

Notice how there are 5 even numbers, and we have to find how many subsets have at most 4 even numbers. Similarly, there are 3 multiples of 3, and we have to find how many elements have at most 2 multiples of 3. We can see that there are fewer cases that don't satisfy the condition, so it is better to use complementary counting. We will first count the number of subsets that have more than 4 even numbers and have more than 2 multiple of 3.

First, how many subsets have more than 4 even numbers?

Since there are 5 multiples of 2 from 1 to 10 $\{2, 4, 6, 8, 10\}$, to have more than 4 multiples of 2 means having all of the multiples of 2 (only 1 way to choose this). From the remaining 5 numbers, there are 2^5 possible subsets. Therefore, in total, there are $1 \times 2^5 = 2^5$ subsets that have more than 4 multiples of 2.

Next, how many subsets have more than 2 multiples of 3?

Since there are 3 multiples of 3 from 1 to 10 $(3, 6, 9)$, to have more than 2 multiples of 3 means having all of the multiples of 3 (only 1 ways to choose this). From the remaining 7 numbers, there are 2^7 possible subsets. By similar logic, in total, there will be $1 \times 2^7 = 2^7$ possible subsets.

For a subset to have all the multiples of 2 and 3, it must contain all the elements from 2, 3, 4, 6, 8, 9, 10 (only 1 way to choose this). From the remaining 3 numbers, there are $1 \times 2^3 = 2^3$ possible subsets.

Therefore, in total, the number of subsets that don't satisfy our original condition is $2^7 + 2^5 - 2^3 = 128 + 32 - 8 = 152$

The total number of possible subsets of the original set of 10 numbers is $2^10 = 1024$ so the number of valid subsets is $1024 - 152 = 872$.

Example 6.4

Mark and George are playing a guessing game. George randomly chooses 2 numbers between 1 and 25. Mark then asks George if exactly 1 of the 2 numbers is divisible by 7. George, truthfully, answers yes. Mark then proceeds to guess 1 multiple of 7 and 1 number that is not a multiple of 7. What is the probability that Mark guesses at least 1 number correctly?

> **Example 6.5**
> How many different ways are there to arrange the letters in the word LOLLIPOP if each of the L's must be next to an O?

§6.3 PIE for 3 Events

> **Theorem 6.6**
> (Principle of Inclusion and Exclusion for 3 Sets)
> $$|A \cup B \cup C| = |A| + |B| + |C| - |A \cap B| - |B \cup C| - |C \cup A| + |A \cup B \cup C|$$
> In this formula, we count the number of possibilities in 3 "things", subtract the possibilities that are duplicates in all 3 pairs of sets, and add back the number of duplicates that are in all 3 sets.

> **Example 6.7**
> How many numbers less than or equal to 193 are multiples of 2, 3, or 5?

First, let's find how many numbers are divisible by 2, how many numbers are divisible by 3, and how many numbers are divisible by 5.

Every 2nd number is divisible by 2, so all the numbers from 2×1 to 2×96 work. Therefore, there are 96 numbers divisible by 2 less than or equal to 193.

Every 3rd number is divisible by 3, so all the numbers from 3×1 to 3×64 work. Therefore, there are 64 numbers divisible by 3 less than or equal to 193

Every 5th number is divisible by 5, so all the numbers from 5×1 to 5×38 work. Therefore, there are 38 numbers divisible by 5 less than or equal to 193.

Next, we must subtract numbers divisible by two of the numbers 2, 3, or 5.

For this to happen, what values must the numbers be divisible by?

Numbers divisible by 2 and 3 or numbers divisible by $2 \times 3 = 6$ are counted twice. Similarly, numbers divisible by 2 and 5 or $2 \times 5 = 10$ are counted twice and numbers divisible by 3 and 5 or $3 \times 5 = 15$ are counted twice. Therefore, we are overcounting multiples of each of 6, 10, and 15, so we must subtract them.

Every 6th number is divisible by 6, so all the numbers from 6×1 to 6×32 work. Therefore, there are 32 numbers divisible by 6 less than or equal to 193.

Every 10th number is divisible by 10, so all the numbers from 10×1 to 10×19 work. Therefore, there are 19 numbers divisible by 10 less than or equal to 193

Every 15th number is divisible by 5, so all the numbers from 15×1 to 15×12 work. Therefore, there are 12 numbers divisible by 15 less than or equal to 193.

If we subtract the overcounted numbers above, did we remove too much?

Next, we look at how many times multiples of 2, 3, and 5, or multiples of $2 \times 3 \times 5 = 60$ were counted. Well, they were originally counted 3 times as part of the multiples of 2, 3, and 5.

Then, they were subtracted 3 times as part of the multiples of 6, 10, and 15. Therefore, they have been counted 0 times!

But we need to count them exactly once, so we must add back the number of multiples of $2 \times 3 \times 5 = 60$. Every 60th number is divisible by 60, so all the numbers from 60×1 to 60×3 work.

Therefore, there are 3 numbers divisible by 60 less than or equal to 193.

In total, to find the number of multiples of 2, 3, or 5 we must add the number of multiples of 2, the number of multiples of 3, the number of multiples of 5. Then, we must subtract the number of multiples of 6, 10, and 15. And finally, we must add back the number of multiples of 60.

This gives us an answer of $96 + 64 + 38 - 32 - 19 - 12 + 3 = 138$.

> **Example 6.8**
> (AMC 8) Five different awards are to be given to three students. Each student will receive at least one award. In how many ways can the awards be distributed?

§6.4 PIE for n Number of Events

The same principle to applies to more than when counting more than 3 events.

> **Theorem 6.9**
> For any n events, to find the total number of ways such that any of the events occur:
>
> 1. Find number of ways each individual event occurs in.
>
> 2. For every pair of 2 events, subtract number of ways where both of them occur
>
> 3. For every triplet of 3 events, add back number of ways where all 3 of them occur
>
> 4. For every quadruplet of 4 events, subtract number of ways where all 4 occur...
>
> 5. For all n events, add (if n is odd) or subtract (if n is even) number of ways where all of them occur
>
> Alternate between adding the number of ways and subtracting the number of ways until you reach the case where all n events occur. Whether you add or subtract the number of ways where all n events occur depends on whether n is odd since we alternate between adding and subtracting.

§6.5 Practice Questions

Essential Academy (June 2023) — Introduction to AMC 8

Problem 6.10
How many positive numbers less than or equal to 100 are multiples of 2 or 3

Problem 6.11
How many positive numbers less than or equal to 100 are NOT multiples of 2 or 3?

Problem 6.12
Out of 200 students, there are 100 taking Geometry, 70 taking Algebra, and 30 taking both. How many students are taking neither?

Problem 6.13
In a town of 351 adults, every adult owns a car, motorcycle, or both. If 331 adults own cars and 45 adults own motorcycles, how many of the car owners do not own a motorcycle?

Problem 6.14
At Euler Middle School, 198 students voted on two issues in a school referendum with the following results: 149 voted in favor of the first issue and 119 voted in favor of the second issue. If there were exactly 29 students who voted against both issues, how many students voted in favor of both issues?

Problem 6.15
How many positive numbers less than or equal to 200 are multiples of 2 and 3 but not a multiple of 5?

Problem 6.16
(AMC 10) There are 20 students participating in an after-school program offering classes in yoga, bridge, and painting. Each student must take at least one of these three classes but may take two or all three. There are 10 students taking yoga, 13 taking bridge, and 9 taking painting. There are 9 students taking at least two classes. How many students are taking all three classes?

Problem 6.17
(AMC 8) Mrs. Sanders has three grandchildren, who call her regularly. One calls her every three days, one calls her every four days, and one calls her every five days. All three called her on December 31, 2016. On how many days during the next year did she not receive a phone call from any of her grandchildren?

Essential Academy (June 2023)   Introduction to AMC 8

Problem 6.18
(AMC 10) Alice refuses to sit next to either Bob or Carla. Derek refuses to sit next to Eric. How many ways are there for the five of them to sit in a row of 5 chairs under these conditions?

Problem 6.19
(AIME) Many states use a sequence of three letters followed by a sequence of three digits as their standard license-plate pattern. Given that each three-letter three-digit arrangement is equally likely, the probability that such a license plate will contain at least one palindrome (a three-letter arrangement or a three-digit arrangement that reads the same left to right as it does right to left) is $\frac{m}{n}$, where m and n are relatively prime positive integers. Find $m+n$.

Problem 6.20
Students are being assigned to faculty mentors in the Berkeley Math Department. If there are 7 distinct students and 3 distinct mentors, and each student has exactly one mentor, in how many ways can students be assigned to mentors such that each mentor has at least one student?

Problem 6.21
(AMC 12) Call a number prime-looking if it is composite but not divisible by 2, 3, or 5. The three smallest prime-looking numbers are 49, 77, and 91. There are 168 prime numbers less than 1000. How many prime-looking numbers are there less than 1000?

Problem 6.22
How many different subsets of $\{1,2,3,4,5,6,7,8,9,10,11,12\}$ contain at least one element in common with each of the sets $\{2,4,6,8,10,12\}$, $\{3,6,9,12\}$ and $\{2,3,5,7,11\}$?

§7 Stars and Bars

§7.1 Stars and Bars Fundamentals

Example 7.1
How many ways are there to distribute 8 identical computers amongst 3 students?

Solution. Make sure to keep in mind that the computers are identical. If the computers were not identical, then we would just have 3 choices for each computer, so our answer would just be 3^8.

However, when the computers are identical, only the number of computers each student has matters. To start, let's say we have 8 computers in a line: $CCCCCCCC$. Therefore, we must divide these computers amongst 3 students.

How can we represent the 8 computers being divided amongst the 3 students?

We can think about having 3 groups. We can simply let the leftmost group be for student one, the middle group for student 2, and the rightmost group for student 3 because only the number of computers each student has matters, so ordering is irrelevant.

How can we divide the 8 C's amongst 3 groups?

We can do this by just placing 2 bars somewhere in between the 8 C's! Everything left of the first bar goes to student 1, everything in the middle goes to student 2, and everything to the right of the 2nd bar goes to student 3. Therefore, the number of ways to distribute the computers is simply the number of ways to insert 2 bars in between 8 C's.

Examples:

- $CC/CCCCC/C$ - 2 for student 1, 5 for student 2, 1 for student 3
- $/CCCC/CCCC$ - 0 for student 1, 4 for student 2, 4 for student 3
- $C/CCCCCCC/$ - 1 for student 1, 7 for student 2, 0 for student 3
- $CCCCCCCC//$ - 8 for student 1, 0 for student 2, 0 for student 3

Now, how do we find the number of ways to insert 2 bars in between 8 C's?

We can think of having 10 slots (each slot for either a bar or a C). Then, we can choose 2 of these slots for bars. Therefore, the numbers of ways to select two slots out of the 10 slots (i.e. choose two slots to put the two bars out of the 10 slots) is $_{10}C_2 = 45$.

Example 7.2
How many ways are there to distribute 8 identical pencils amongst 3 people?

> **Theorem 7.3**
> (Stars and Bars) The number of ways to place n indistinguishable objects into k distinguishable boxes is
> $$_{n+k-1}C_n$$

§7.2 Stars and Bars With Constraints

Stars and bars is extremely useful, and can often be adapted based on situations. For example, if each bin has to have at least 1 object in it we assign each bin 1 object to start off with and apply our formula with $n - k$ objects and k distinguishable boxes.

> **Example 7.4**
> 7 astronauts are stranded in space with only 10 identical meals left. Every astronaut must receive at least 1 meal or else, they will starve. How many ways are there to distribute meals under this constraint?

Solution. **How can we deal with the condition that every astronaut must get at least 1 meal?**

If every astronaut must receive at least 1 meal and all the meals are identical, we can simply pre-distribute 7 of the meals and give 1 to each astronaut. Notice that now, no matter where the remaining meals are distributed, every astronaut will get 1 meal or more.

Now, how can we use stars on bars on the remaining problem?

After 7 meals are distributed, only 3 remain. There is no constraint on who these 3 meals can go to, so it can go to any of the 7 astronauts. To divide these 3 meals into 7 groups, we can place 6 bars in between the 3 meals. Then, out of the $3 + 6 = 9$ slots for meals and bars, we must choose 6 of them for bars. So our answer is $_9C_6 = 84$.

> **Example 7.5**
> What's the probability that the sum of the top faces on five regular 6-sided dice is 24?

Solution. Before we begin, note that each dice must be a value from 1 to 6. There is no way to use stars and bars directly because it will not account for the constraint that the maximum of each dice roll is a 6.

How can we use stars and bars for this problem?

Let the default for each dice be a 6. Then the maximum score is 30, we must essentially distribute 6 negatives to the 5 dice $(-1, -1, -1, -1, -1, -1)$. We can see that to distribute 6 negatives to 5 dice, we must have 4 bars to separate the negatives into 5 groups for the dice. We then have 9 slots and must choose 4 of them for bars. We can do this in $_9C_4 = 126$.

Are there any cases we are overcounting?

Because the minimum value of a dice roll is 1, each dice can only take at most 5 negatives. Therefore, we are overcounting the possibilities where 1 dice receives 6 negatives.

How many ways are there for 1 dice to have 6 negatives?

There are 6 dice, so there are 6 choices for which dice will have all 6 negatives. From here, we can calculate our final answer by subtracting the 6 ways that result in 1 dice receiving all 6 negatives from $_9C_4$, the number of ways to distribute 6 negatives to 5 dice so that no dice gets more than 5 negatives.

Therefore $_9C_5 - 6 = 120$.

Example 7.6
You have 11 Peaches and 5 bananas. You place them into 3 baskets. How many ways are there to do this if each basket must have more peaches than bananas and at least 1 of any fruit?

§7.3 Practice Questions

Problem 7.7
How many ways are there to distribute 5 identical candies to 3 children?

Problem 7.8
(AMC 10) Pat is to select six cookies from a tray containing only chocolate chip, oatmeal, and peanut butter cookies. There are at least six of each of these three kinds of cookies on the tray. How many different assortments of six cookies can be selected?

Problem 7.9
(AMC 8) Alice has 24 apples. In how many ways can she share them with Becky and Chris so that each of the three people has at least two apples?

Problem 7.10
There are six different choices on the dollar menu at Burger Bar: fries, apple pies, onion rings, chicken nuggets, hamburgers, and cheeseburgers. You decide to purchase four of these items. How many different combinations of four items can you buy from the dollar menu?

Problem 7.11
You are purchasing a dozen roses for Valentine's Day. The roses come in red, white, and pink and you want at least one of each color. How many different bouquets of one dozen roses are possible?

Problem 7.12
You are playing a racing video game. To begin, you get to adjust the tuning of your race car by adding a combined total of ten points to three categories. You can adjust your speed, handling, and acceleration by adding anywhere from 0 to 10 points to each category. How many tuning options are there for the car's initial setup?

Problem 7.13
A standard six-faced die is rolled four times. How many ways are there to roll a 9?

§8 Geometric Counting

§8.1 Geometric Counting Fundamentals

Example 8.1
How many triangles can be formed by connecting vertices in a regular octagon?

Solution. Octagon has 8 vertices. Therefore, our answer is $_8C_3 = 56$.

§8.2 Number of Squares in Grid

Example 8.2
How many squares of all sizes can be formed from a 4×4 grid of unit squares?

Solution.

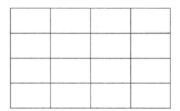

What's the easiest and most systematic way to count all possible squares?

We can break the problem down into $1 \times 1, 2 \times 2, 3 \times 3$, and 4×4 squares.

Case 1: 1×1 squares For this case, there are just $4 \times 4 = 16$ squares.

Case 2: 2×2 squares Notice how there are 3 possible choices of rows and 3 possible choices of columns to form a 2×2 square. In total, there are $3 \times 3 = 9$ possible 2×2 squares since each choice of two rows and two columns produces a 2×2 square.

Case 3: 3×3 squares

Notice how there are 2 possible choices of rows and 2 possible choices of columns to form a 3×3 square. In total, there are $2 \times 2 = 4$ possible 3×3 squares since each choice of two rows and two columns produces a 3×3 square.

Case 4: 4×4 squares

There is just one 4×4 square (the entire square). In total, the number of squares of all sizes is $16 + 9 + 4 + 1 = 30$

Example 8.3
The number of squares in a $n \times n$ grid of squares is
$$1^2 + 2^2 + \cdots + n^2 = \frac{n(n+1)(2n+1)}{6}$$

§8.3 Number of Rectangles in Grid

Example 8.4
How many rectangles are in a 3×5 grid of rectangles?

Solution. **What defines a rectangle in a grid?**

4 points? Not quite because 4 points don't necessarily make a rectangle. 2 vertical lines and 2 horizontal lines define a rectangle instead. In other words, any 2 vertical lines and horizontal lines in a grid will form a rectangle.

How can we count the number of rectangles in a grid based on this? We simply have to choose 2 vertical lines and 2 horizontal lines! There are 4 horizontal lines, so there are $_4C_2$ ways to select 2 horizontal lines that define a rectangle. Similarly, there are 6 vertical lines, so there are $_6C_2$ ways to select 2 vertical lines that define a rectangle. Therefore, the number of rectangles is just $_4C_2 \times _6C_2 = 90$.

Lemma 8.5
The general formula for the number of rectangles of all sizes in a rectangular grid of size $m \times n$ is
$$_{m+1}C_2 \times _{n+1}C_2$$

Example 8.6
How many rectangles, which are not squares, are in a 5×4 grid of squares?

Example 8.7
How many squares with side length $\sqrt{5}$ can be found in a 5 by 5 grid of points?

§8.4 Path Counting

Example 8.8
A rabbit is walking in a coordinate plane and must get from $(0,0)$ to the carrot at the point $(5,5)$. If the rabbit can only move 1 unit up or 1 unit right on any given move, how many ways are there for the rabbit to get to the carrot?

Solution. Let's find a simpler way to solve this problem. Lets denote a right move by R and an up move by U.

How many right and up moves does the rabbit need to get to the carrot?

Because it's going from $(0,0)$ to $(5,5)$, it needs 5 right moves and 5 up moves. Therefore, we just need to count the number of ways to make 5 right moves and 5 up moves in any order.

How can we represent the number of ways to make 5 right moves and 5 up moves (in any order)?

This can be thought of as rearrangements of $5R$ and $5U$: $RRRRRUUUUU$. We just need to find all possible rearrangements of this word. Given the total number of characters is 10, and counting the duplicates ($5R$'s and $5U$'s), we can use the word rearrangement formula:

$$\frac{10!}{5!5!} = 252$$

Example 8.9
How many ways to go from $(0,0)$ to $(5,6)$ moving only right or up if you cannot visit all of $(1,2), (3,4)$, and $(5,5)$ because there is a monster that will eat you if you visit all 3 locations?

Example 8.10
(BMMT) Sally is inside a pen consisting of points (a,b) such that $0 \le a \le b \le 4$. If she is currently on the point (x,y), she can move to either $(x, y+1), (x, y-1)$, or $(x+1, y)$. Given that she cannot revisit any point she has visited before, find the number of ways she can reach $(4,4)$ from $(0,0)$.

Example 8.11

(AMC 8) A game board consists of 64 squares that alternate in color between black and white. The figure below shows square P in the bottom row and square Q in the top row. A marker is placed at P. A step consists of moving the marker onto one of the adjoining white squares in the row above. How many 7-step paths are there from P to Q? (The figure shows a sample path.)

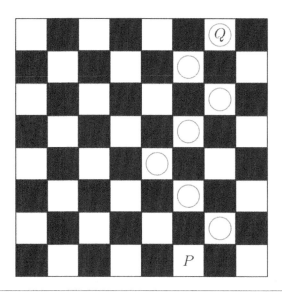

§8.5 Practice Problems

Problem 8.12
How many rectangles of any size are in a 4×6 grid of squares?

Problem 8.13
A mouse is standing on a grid at location $(0,0)$. There is cheese on the grid at $(5,5)$. He can only move right or up. How many different paths can he take to the cheese?

Problem 8.14
Samantha lives 2 blocks west and 1 block south of the southwest corner of City Park. Her school is 2 blocks east and 2 blocks north of the northeast corner of City Park. On school days she bikes on streets to the southwest corner of City Park, then takes a diagonal path through the park to the northeast corner, and then bikes on streets to school. If her route is as short as possible, how many different routes can she take?

Problem 8.15
(BMMT) Leanne and Jing Jing are walking around the xy-plane. In one step, Leanne can move from any point (x,y) to $(x+1,y)$ or $(x,y+1)$ and Jing Jing can move from (x,y) to $(x-2,y+5)$ or $(x+3,y-1)$. The number of ways that Leanne can move from $(0,0)$ to $(20,20)$ is equal to the number of ways that Jing Jing can move from $(0,0)$ to (a,b), where a and b are positive integers. Compute the minimum possible value of $a+b$.

§9 Recursion

§9.1 Recursion Fundamentals

Recursion is the process of finding smaller values and using them to calculate larger values.

> **Example 9.1**
> Jongwook is climbing a 6 stair staircase. If he can climb 1 or 2 stairs at a time, in how many distinct ways can he get to the top?

Solution. We could do casework, but that would take a long time. Instead, let's solve this problem for small values first.

How many ways are there for Jongwook to reach the 0th stair (ground)?

Jongwook is already there at the start, so there is just 1 way to do that.

How many ways are there for Jongwook to reach the 1st stair?

There is only 1 way because he must climb 1 step to reach there from the ground (0th stair).

How many ways are there for Jongwook to reach the 2nd stair?

There are 2 ways: 2 single steps or 1 double step from ground (0th stair)

How can Jongwook reach the 3rd stair?

Jongwook either has to take a single step from the 2nd stair, or a double step from the 1st stair. Therefore, the number of ways is simply the number of ways to reach the first step plus the number of ways to reach the 2nd step, which is $1 + 2 = 3$.

How can use the same logic to find the number of ways to reach the nth stair?

To reach the nth stair, Jeff must take a 1 step from the $(n-1)^{\text{th}}$ stair or a double step from the $(n-2)^{\text{th}}$ stair. Therefore, the number of ways of reaching the nth step is the number of ways of reaching the $(n-2)^{\text{th}}$ plus the number of ways of reaching the $(n-2)^{\text{th}}$ stair.

Let's define a function, $f(n)$, such that $f(n)$ represents the number of ways to reach the nth step. Then, we can write the recursion $f(n) = f(n-1) + f(n-2)$. The next step is to just iteratively calculate the values of $f(n)$ for all values of n until we find $f(6)$, since the staircase has 6 steps. We already calculated $f(1)$ and $f(2)$ so we can use them as base cases.

- $f(1) = 1$
- $f(2) = 2$
- $f(3) = f(3-1) + f(3-2) = f(2) + f(1) = 2 + 1 = 3$
- $f(4) = f(4-1) + f(4-2) = f(3) + f(2) = 3 + 2 = 5$
- $f(5) = f(5-1) + f(5-2) = f(4) + f(3) = 5 + 3 = 8$

- $f(6) = f(6-1) + f(6-2) = f(5) + f(4) = 8 + 5 = 13$

§9.2 Recursion with Constraints

> **Example 9.2**
> Mike is climbing a staircase with 10 stairs. He is in a rush, so he will climb 2 or 3 steps at a time for most of the journey. If he reaches the 9th stair, he is allowed to climb 1 stair to reach the 10th stair. How many ways are there for Mike to reach the top of the staircase?

Solution. We can do a similar approach to the previous problem. We first begin by finding the base cases. There is no way to reach the 1st stair since he cannot take a 1 step, so $f(1) = 0$. There is 1 way to reach the 2nd stair by taking a double step, so $f(2) = 1$. There is 1 way to reach the 3rd stair by taking a triple step. If the first step is a double step, then it's impossible to reach the 3rd stair since there are no single steps. Therefore, $f(3) = 1$.

Next, what is the recursive function for this problem?

To get the the nth stair, we must either take a 2 step from the $(n-2)^{\text{th}}$ stair or a 3 step from the $(n-3)^{\text{th}}$ stair.

So, the recursion relation would be

$$f(n) = f(n-2) + f(n-3)$$

Let's now evaluate the values of $f(n)$

- $f(1) = 0$
- $f(2) = 1$
- $f(3) = 1$
- $f(4) = f(2) + f(1) = 1$
- $f(5) = f(3) + f(2) = 2$
- $f(6) = f(4) + f(3) = 2$
- $f(7) = f(5) + f(4) = 3$
- $f(8) = f(6) + f(5) = 4$
- $f(9) = f(7) + f(6) = 5$
- $f(10) = f(8) + f(7) = 7$

So, is our answer just $f(10) = 7$?

No, because once he reaches stair 9, he is allowed to take a single step. So, we must add the number of ways to reach stair 9 to the number of ways to reach stair 8 and the number of ways to reach stair 7. Therefore, we must find: $f(10) + f(9) = 12$.

§9.3 Probability Recursion

Example 9.3
A cricket randomly hops between 4 leaves, on each turn hopping to one of the other 3 leaves with equal probability. After 4 hops what is the probability that the cricket has returned to the leaf where it started?

§9.4 Practice Problems

Problem 9.4
(AMC 8) Everyday at school, Jo climbs a flight of 6 stairs. Jo can take the stairs 1, 2, or 3 at a time. For example, Jo could climb 3, then 1, then 2. In how many ways can Jo climb the stairs?

Problem 9.5
(AMC 12) Call a set of integers *spacy* if it contains no more than one out of any three consecutive integers. How many subsets of $\{1, 2, 3, \ldots, 12\}$, including the empty set, are *spacy*?

Problem 9.6
(AIME) A collection of 8 cubes consists of one cube with edge-length k for each integer k, $1 \leq k \leq 8$. A tower is to be built using all 8 cubes according to the following rules: Any cube may be the bottom cube in the tower. The cube immediately on top of a cube with edge-length k must have edge-length at most $k + 2$. How many towers can be constructed?

Problem 9.7
What is the probability of selecting a heart from a shuffled deck of cards?

Problem 9.8
What is the probability of selecting three red cards in a row with replacement?

Problem 9.9
What is the probability of selecting two cards from different suits with replacement?

Problem 9.10
What is the probability that the top two cards of a shuffled deck are both face cards? (Kings, Queens, and Jacks are all face cards.)

Problem 9.11
What is the probability that the top four cards of a shuffled deck are all of the same suit?

Problem 9.12
What is the probability that the top card in a shuffled deck is a red Ace, and second card is a spade?

Problem 9.13
What is the probability that the top card in a shuffled deck is an Ace(of any color), and the second card is a spade? The Ace may be the Ace of spades.

Problem 9.14
What is the probability that the top two cards in a shuffled deck are an Ace (of any suit) and a spade? The Ace of spades cannot be used to satisfy both requirements.

Problem 9.15
What is the probability that the top two cards in a shuffled deck are consecutive cards of the same suit? (The Ace can be high or low).

Problem 9.16
A fair coin is flipped three times, and each flip comes up heads. What is the probability that the next flip will also be heads?

Problem 9.17
You roll a pair of dice; one red and the other green. What is the probability of rolling a five on the red die and an even number on the green die?

Problem 9.18
For a lottery drawing, a set of balls numbered 1 through 9 is placed in each of three bins. One ball is selected from each bin. What is the probability that all three digits drawn will be odd?

Problem 9.19
For a randomly selected phone number, what is the probability that the last three digits are all the same?

Problem 9.20
What is the probability of rolling each of the numbers 1 through 6 in any order in six rolls of a standard die?

Problem 9.21
A dot is added to each face of two standard dice. What is the probability of rolling an 8 with the modified pair of dice?

Problem 9.22
Six students are randomly seated in the first row of a classroom. What is the probability that from left to right they are in order from oldest to youngest?

Problem 9.23
Each spinner below is divided into four equal sectors. Each is spun once, and the product of the spins is calculated. What is the probability that the product is positive?

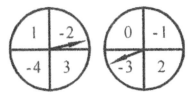

Problem 9.24
For three randomly selected positive integers, what is the probability that the sum of the units digits is even?

Algebra

§10 Ratio and Percentage

§10.1 Ratio Fundamentals

Example 10.1
(MATHCOUNTS) The Ten Finger calculator company periodically checks random calculators before shipping crates out to customers. On Wednesday, 12 calculators from each of 64 crates of 144 calculators were tested. Two of the tested calculators were found to be defective. Based on this rate of defect, how many total calculators are expected to be defective?

Example 10.2
Alan has a jar with many marbles. 20% of them are small, 40% are medium, and 40% are large. For each size, 30% of the marbles are red, 20% blue, 40% green, and 10% yellow. Half of large yellow and small green marbles have a special design on them. What percentage of the total marbles have a special design?

Example 10.3
(AMC 10) Two jars each contain the same number of marbles, and every marble is either blue or green. In Jar 1 the ratio of blue to green marbles is $9:1$, and the ratio of blue to green marbles in Jar 2 is $8:1$. There are 95 green marbles in all. How many more blue marbles are in Jar 1 than in Jar 2?

Example 10.4
There are 994 students in the Essential Middle School. $\frac{2}{7}$ of them are in 6th grade, $\frac{2}{7}$ of them are in 7th grade, and $\frac{3}{7}$ of them are in 8th grade. $\frac{1}{4}$ of the 6th graders are interested in math, $\frac{3}{4}$ of the 7th graders are interested in math, and $\frac{1}{2}$ of the 8th graders are interested in math. Of the students who are interested in math, $\frac{5}{7}$ will take the AMC 8 this year. How many students will take the AMC 8 from Essential Middle School?

§10.2 Rate and Work

Definition 10.5. (Relations Between Work, Rate, and Time)

$$\text{Amount of Work} = \text{Rate} \times \text{Time}$$

$$\text{Rate} = \frac{\text{Amount of Work}}{\text{Time}}$$

$$\text{Time} = \frac{\text{Amount of Work}}{\text{Time}}$$

Example 10.6

10 workers from a company each working at a constant rate can build 10 houses in 12 years. However, after 6 years, 5 of them retire. The other workers continue working at the same original rate. The project is falling behind, so 2 years later, the company hires 10 more workers who each work twice as fast as the original workers. After how many total years will the 10 houses be complete?

Solution. **After 6 years, what fraction of the 10 houses will be built?**

Since 6 years is half of the time needed to build 10 houses, $\frac{1}{2}$ of the houses will be completed by 6 years. Next, we can see that for 2 years, only 5 of the original workers are working.

In these 2 years, what fraction of the houses will be built?

2 years is $\frac{1}{6}$ of the time needed to build all of the houses. However, only $\frac{1}{2}$ of the original workers are working. Therefore, in total, $\frac{1}{2} \times \frac{1}{6} = \frac{1}{12}$ of the houses will be built in these 2 years.

Next, we must find the rate of building houses when the 10 workers are added.

Compared to the original rate, how fast do the 5 original and 10 new workers work?

Since each of the 10 new workers work 2 times faster than one of the original workers, the 5 original workers that remain and the 10 new workers can do as much work as $5 + 10 \times 2 = 25$ original workers. Therefore, with the addition of the new workers, they can work $\frac{25}{10} = \frac{5}{2}$ times faster.

What fraction of the houses are left to be built?

$\frac{1}{2}$ of the houses were built in the first 6 years, and $\frac{1}{12}$ of the houses were built in the next 2 years, so

$$1 - \frac{1}{2} - \frac{1}{12} = \frac{5}{12}$$

of the houses are left to be built.

How long would the 10 original workers have taken to build $\frac{5}{12}$ of the houses?

Since they could build all of the houses in 12 years, they could build $\frac{5}{12}$ of the houses in $\frac{5}{12} \times 12 = 5$ years.

How long will it take the new team of workers to build $\frac{5}{12}$ of the houses?

The new team of workers can work $\frac{5}{2}$ times faster than the original team, so they will take $\frac{2}{5}$ the time. Therefore, they can build $\frac{5}{12}$ of the houses in $\frac{2}{5} \times 5 = 2$ years.

The 10 original workers built $\frac{1}{2}$ of the houses in 6 years, the remaining 5 original workers built $\frac{1}{12}$ of the houses in 2 years, and the new team of workers built $\frac{5}{12}$ of the houses in 2 years. Therefore, the total number of years it took to build the houses is $6 + 2 + 2 = 10$.

Remark 10.7. The if one person can do something is a amount of time, and someone else can do it in b amount of time, together they can do it in

$$\frac{ab}{a+b}$$

amount of time.

Example 10.8
Alex, using substitution or elimination, can solve 20 two variable equations in 10 minutes. Bob, who uses the diagonal product method, can solve 20 two variable equations in 2 minutes. How much time will it take both of them working together to solve 60 two variable equations?

Example 10.9
All experienced workers work at a constant rate and all new workers work at a different constant rate. 12 experienced workers and 6 new workers can build a house in 6 months. 6 experienced workers and 12 new workers can build a house in 9 months. How many months will it take 9 experienced workers and 9 new workers to build a house?

Example 10.10
Two rocket scientists Einstein and Oppenheimer are building a spaceship, and they will work 6 hours a day for standard pay. It would take Oppenheimer a total of 48 hours to build the rocket individually, and it would take Einstein a total of 36 hours. The manager can motivate any of them to work extra hours every day by giving them an extra $100 for each hour. The manager only has $1000 to give to Oppenheimer and Einstein for extra pay. What is the minimum number of whole days that the rocket can be completed?

§10.3 Practice Problems

Problem 10.11
If the degree measures of the angles of a triangle are in the ratio $3 : 3 : 4$, what is the degree measure of the largest angle of the triangle?

Problem 10.12
There are 270 students at Colfax Middle School, where the ratio of boys to girls is $5 : 4$. There are 180 students at Winthrop Middle School, where the ratio of boys to girls is $4 : 5$. The two schools hold a dance and all students from both schools attend. What fraction of the students at the dance are girls?

Problem 10.13
Gilda has a bag of marbles. She gives 20% of them to her friend Pedro. Then Gilda gives 10% of what is left to another friend, Ebony. Finally, Gilda gives 25% of what is now left in the bag to her brother Jimmy. What percentage of her original bag of marbles does Gilda have left for herself?

Problem 10.14

All of Marcy's marbles are blue, red, green, or yellow. One third of her marbles are blue, one fourth of them are red, and six of them are green. What is the smallest number of yellow marbles Marcy can have?

Problem 10.15

A number of students from Fibonacci Middle School are taking part in a community service project. The ratio of 8th-graders to 6th-graders is $5:3$, and the ratio of 8th-graders to 7th-graders is $8:5$. What is the smallest number of students that could be participating in the project?

Problem 10.16

Chloe and Zoe are both students in Ms. Demeanor's math class. Last night they each solved half of the problems in their homework assignment alone and then solved the other half together. Chloe had correct answers to only 80% of the problems she solved alone, but overall 88% of her answers were correct. Zoe had correct answers to 90% of the problems she solved alone. What was Zoe's overall percentage of correct answers?

Problem 10.17

Suppose 15% of x equals 20% of y. What percentage of x is y?

Problem 10.18

On Wednesday, $\frac{1}{3}$ of the students in Mr. Short's homeroom had drama practice, $\frac{1}{4}$ of his other homeroom students had band practice. If 6 students had band practice, how many students are in Mr. Short's homeroom?

Problem 10.19

The ratio of w to x is $4:3$, the ratio of y to z is $3:2$, and the ratio of z to x is $1:6$. What is the ratio of w to y?

Problem 10.20

A store increased the original price of a shirt by a certain percent and then decreased the new price by the same amount. Given that the resulting price was 84% of the original price, by what percent was the price increased and decreased?

Problem 10.21

Andre can complete $\frac{5}{6}$ of a job in $\frac{3}{4}$ of the time that it takes Michael to do the whole job. What is the ratio of the rate at which Andre works to the rate at which Michael works? Express your answer as a common fraction.

Problem 10.22
When the World Wide Web first became popular in the 1990s, download speeds reached a maximum of about 56 kilobits per second. Approximately how many minutes would the download of a 4.2-megabyte song have taken at that speed? (Note that there are 8000 kilobytes in a megabyte.)

Problem 10.23
Two jars each contain the same number of marbles, and every marble is either blue or green. In Jar 1 the ratio of blue to green marbles is $9:1$, and the ratio of blue to green marbles in Jar 2 is $8:1$. There are 95 green marbles in all. How many more blue marbles are in Jar 1 than in Jar 2?

Problem 10.24
Grapes are 80% water by weight. When a bushel of grapes dries in the sun for two weeks, it loses 50% of its total weight. All of the weight loss is due to the loss of water. After drying for two weeks, what percentage of the grapes is water by weight? Express your answer to the nearest percent.

Problem 10.25
Brian has earned 65%, 80% and 92% on his three pre-final exams. These exams are not weighed equally: the lowest counts for only 20% of his overall grade, while the other two count for 25% each. If the final exam is the remainder of the overall grade and there are no opportunities for extra credit, what is the highest-grade Brian can earn in the class? Express your answer to the nearest whole percent.

§11 Linear Function and Quadratic Function

§11.1 Linear Function and Quadratic Function

Linear Function

Standard Form:

Slope Intercept Form:

Point Intercept Form:

Quadratic Function

Standard Form

Vertex Form

How to Factor?

Vieta's Formulas

Theorem 11.1

The solutions to the quadratic equation $ax^2 + bx + c = 0$ is
$$x = \frac{-b \pm \sqrt{b^2 - 4ac}}{2a}$$

Example 11.2

$$\begin{cases} 2x + y + z &= 36 \\ x + 2y + z &= 40 \\ x + y + 2z &= 32 \end{cases}$$

Solution. Let's rewrite the equations as follows:

$$\begin{cases} x + (x + y + z) &= 36 \\ y + (x + y + z) &= 40 \\ z + (x + y + z) &= 32 \end{cases}$$

Notice how each equation simply consists of the sum $x + y + z$ plus an additional $x, y,$ or z. The sum $x + y + z$ reoccurs in every equation.

How can we make another $x + y + z$ term?

We can add all 3 equations together! This will give us

$$(x+y+z) + 3(x+y+z) = 4(x+y+z) = 36 + 40 + 32 = 108.$$

This simplifies to $x+y+z = 27$. Therefore, $x = 9, y = 13, z = 5$.

Example 11.3

(EMCC) Suppose x, y, z are real numbers that satisfy

$$\begin{cases} x+y-z = 5 \\ y+z-x = 7 \\ z+x-y = 9 \end{cases}$$

Find $x^2 + y^2 + z^2$.

Example 11.4

(EMCC) Suppose x, y, z are real numbers that satisfy:

$$\begin{cases} x+y-z = 5 \\ y+z-x = 7 \\ z+x-y = 9 \end{cases}$$

Compute $y^2 - x^2$.

Example 11.5

(BMMT) Let x and y be real numbers such that $xy = 4$ and $x^2y + xy^2 = 25$. Find the value of $x^3y + x^2y^2 + xy^3$.

---

§11.2 Word Problems

Example 11.6

Orangey loves drinking diluted juice. Orangey takes an 8-ounce glass of orange juice and drinks some fraction of it. Then, he fills the rest with apple juice. After thoroughly mixing the glass, he drinks two thirds of the mixture. He then fills the rest with water. Orangey then finishes the whole glass. If he drank 4 times as much orange juice than apple juice, how much total liquid in ounces did he drink?

Example 11.7

Jar A contains four liters of a solution that is 45% acid. Jar B contains five liters of a solution that is 48% acid. Jar C contains one liter of a solution that is k% acid. From jar C, $\frac{m}{n}$ liters of the solution is added to jar A, and the remainder of the solution in jar C is added to jar B. At the end both jar A and jar B contain solutions that are 50% acid. Given that m and n are relatively prime positive integers, find $k + m + n$.

§11.3 Practice Problems

Problem 11.8
Shauna takes five tests, each worth a maximum of 100 points. Her scores on the first three tests are 76, 94, and 87. In order to average 81 for all five tests, what is the lowest score she could earn on one of the other two tests?

Problem 11.9
In a jar of red, green, and blue marbles, all but 6 are red marbles, all but 8 are green, and all but 4 are blue. How many marbles are in the jar?

Problem 11.10
$$\begin{cases} abc &= 4 \\ bcd &= 6 \\ cde &= 12 \\ def &= 18 \\ eab &= 9 \end{cases}$$
Find $abcdef$.

Problem 11.11
In a mathematics contest with ten problems, a student gains 5 points for a correct answer and loses 2 points for an incorrect answer. If Olivia answered every problem and her score was 29, how many correct answers did she have?

Problem 11.12
I am thinking of 3 integers. Added two at a time their sums are 37, 41, 44. What is the product of the 3 integers?

Problem 11.13
Hui is an avid reader. She bought a copy of the best seller Math is Beautiful. On the first day, Hui read of the pages plus more, and on the second day she read of the remaining pages plus pages. On the third day she read of the remaining pages plus pages. She then realized that there were only pages left to read, which she read the next day. How many pages are in this book?

Problem 11.14
Before the district play, the Unicorns had won 45% of their basketball games. During district play, they won six more games and lost two, to finish the season having won half their games. How many games did the Unicorns play in all?

Problem 11.15
Ralph went to the store and bought 12 pairs of socks for a total of 24. Some of the socks he bought cost 1 a pair, some of the socks he bought cost 3 a pair, and some of the socks he bought cost 4 a pair. If he bought at least one pair of each type, how many pairs of 1 socks did Ralph buy?

Problem 11.16
Suppose that $a \circ b = 3a - b$. What is the value of x if $2 \circ (5 \circ x) = 1$.

Problem 11.17
In an after-school program for juniors and seniors there is a debate team with an equal number of students from each class on the team. Among the 28 students on the program, 25% of the juniors and 10% of the seniors are on the debate team. How many juniors are in the program?

Problem 11.18
Joe has a collection of 23 coins, consisting of 5-cent coins, 10-cent coins, and 25-cent coins. He has 3 more 10-cent coins than 5-cent coins, and the total value of his collection is 320 cents. How many more 25-cent coins does Joe have than 5-cent coins?

Problem 11.19
Pablo, Sofia, and Mia got some candy eggs at a party. Pablo had three times as many eggs as Sofia, and Sofia had twice as many eggs as Mia. Pablo decides to give some of his eggs to Sofia and Mia so that all three will have the same number of eggs. What fraction of his eggs should Pablo give to Sofia?

Problem 11.20
The sum of two natural numbers is $17,402$. One of the two numbers is divisible by 10. If the units digit of that number is erased, the other number is obtained. What is the difference of these two numbers?

Problem 11.21
Let $a + 1 = b + 2 = c + 3 = d + 4 = a + b + c + d + 5$. What is $a + b + c + d$?

Problem 11.22
$$\begin{cases} a + 5b + 9c = 1 \\ 4a + 2b + 3c = 2 \\ 7a + 8b + 6c = 9 \end{cases}$$
What is $741a + 825b + 639c$?

Problem 11.23
The units and tens digits of one two-digit integer are the tens and units digits of another two-digit integer, respectively. If the product of the two integers is 4930, what is their sum?

Problem 11.24
A car passes point A driving at a constant rate of 60 km per hour. A second car, traveling at a constant rate of 75 km per hour, passes the same point A a while later and then follows the first car. It catches the first car after traveling a distance of 75 km past point A. How many minutes after the first car passed point A did the second car pass point A?

Problem 11.25
Four numbers are written in a row. The average of the first two is 21, the average of the middle two is 26, and the average of the last two is 30. What is the average of the first and last of the numbers?

Problem 11.26
Ana and Bonita were born on the same date in different years, n years apart. Last year Ana was 5 times as old as Bonita. This year Ana's age is the square of Bonita's age. What is n?

Problem 11.27
On the last day of school, Mrs. Awesome gave jelly beans to her class. She gave each boy as many jelly beans as there were boys in the class. She gave each girl as many jelly beans as there were girls in the class. She brought 400 jelly beans, and when she finished, she had six jelly beans left. There were two more boys than girls in her class. How many students were in her class?

Problem 11.28
At Megapolis Hospital one year, multiple-birth statistics were as follows: Sets of twins, triplets, and quadruplets accounted for 1000 of the babies born. There were four times as many sets of triplets as sets of quadruplets, and there was three times as many sets of twins as sets of triplets. How many of these 1000 babies were in sets of quadruplets?

Problem 11.29
Two jars each contain the same number of marbles, and every marble is either blue or green. In Jar 1 the ratio of blue to green marbles is $9:1$, and the ratio of blue to green marbles in Jar 2 is $8:1$. There are 95 green marbles in all. How many more blue marbles are in Jar 1 than in Jar 2?

Problem 11.30
The product of three distinct positive integers is 144. If the sum of the three integers is 26, what is the sum of their squares?

Essential Academy (June 2023)   Introduction to AMC 8

Problem 11.31

The state income tax where Kristin lives is levied at the rate of $p\%$ of the first 28000 of annual income plus $(p+2)\%$ of any amount above 28000. Kristin noticed that the state income tax she paid amounted to $(p+0.25)\%$ of her annual income. What was her annual income?

Problem 11.32

Paula the painter and her two helpers each paint at constant, but different, rates. They always start at 8 : 00 AM, and all three always take the same amount of time to eat lunch. On Monday the three of them painted 50% of a house, quitting at 4 : 00 PM. On Tuesday, when Paula wasn't there, the two helpers painted only 24% of the house and quit at 2 : 12 PM. On Wednesday Paula worked by herself and finished the house by working until 7 : 12 P.M. How long, in minutes, was each day's lunch break?

§12 Speed, Distance, and Time

§12.1 Speed, Distance and Time

Definition 12.1.
$$\text{Distance} = \text{Speed} \times \text{Time}$$
$$\text{Speed} = \frac{\text{Distance}}{\text{Time}}$$
$$\text{Time} = \frac{\text{Distance}}{\text{Speed}}$$

Example 12.2

Messi needs to get to his office 40 miles away. If he drives at 60 miles per hour, how many minutes will it take him to get to work?

Example 12.3

Ronaldo is running along a 2 mile path at 8 miles per hour and back on the same path. On the way back he is tired so for after every 1 mile he runs on the way back, his speed is instantly reduced by 1 mile per hour. What is his average speed throughout the whole trip?

Example 12.4

(AMC 8) Each day for four days, Linda traveled for one hour at a speed that resulted in her traveling one mile in an integer number of minutes. Each day after the first, her speed decreased so that the number of minutes to travel one mile increased by 5 minutes over the preceding day. Each of the four days, her distance traveled was also an integer number of miles. What was the total number of miles for the four trips?

Example 12.5

(AMC 8) A bus takes 2 minutes to drive from one stop to the next, and waits 1 minute at each stop to let passengers board. Zia takes 5 minutes to walk from one bus stop to the next. As Zia reaches a bus stop, if the bus is at the previous stop or has already left the previous stop, then she will wait for the bus. Otherwise she will start walking toward the next stop. Suppose the bus and Zia start at the same time toward the library, with the bus 3 stops behind. After how many minutes will Zia board the bus?

Example 12.6

(AIME) Points $A, B,$ and C lie in that order along a straight path where the distance from A to C is 1800 meters. Ina runs twice as fast as Eve, and Paul runs twice as fast as Ina. The three runners start running at the same time with Ina starting at A and running toward C, Paul starting at B and running toward C, and Eve starting at C and running toward A. When Paul meets Eve, he turns around and runs toward A. Paul and Ina both arrive at B at the same time. Find the number of meters from A to B.

§12.2 Harmonic Mean

Definition 12.7. The **harmonic mean** of a and b is defined to be

$$\frac{2}{\frac{1}{a}+\frac{1}{b}} = \frac{2ab}{a+b}$$

Example 12.8

Kelly rides her bicycle again up a hill and then back home. On the way up she averages a miles per hour, and on the way home she averages b miles per hour. What is his average speed for the entire trip in terms of a and b?

Solution. If we use ab as the length of the hill, it will take Kelly $\frac{ab}{a} = b$ hours to ride up the hill and it will take $\frac{ab}{b} = a$ hours to ride back down the hill. The total distance traveled will be $2ab$ miles and the total time will be $a+b$ hours. This gives us his average speed for the entire trip, called the harmonic mean of a and b.

Example 12.9

What is Ariana's average speed on her walk to school if she walks halfway at 2mph and runs the rest of the way at 7mph?

---

§12.3 Practice Problems

Problem 12.10

Qiang drives 15 miles at an average speed of 30 miles per hour. How many additional miles will he have to drive at 55 miles per hour to average 50 miles per hour for the entire trip?

Problem 12.11

Chantal and Jean start hiking from a trailhead toward a fire tower. Jean is wearing a heavy backpack and walks slower. Chantal starts walking at 4 miles per hour. Halfway to the tower, the trail becomes really steep, and Chantal slows down to 2 miles per hour. After reaching the tower, she immediately turns around and descends the steep part of the trail at 3 miles per hour. She meets Jean at the halfway point. What was Jean's average speed, in miles per hour, until they meet?

Problem 12.12

Bella begins to walk from her house toward her friend Ella's house. At the same time, Ella begins to ride her bicycle toward Bella's house. They each maintain a constant speed, and Ella rides 5 times as fast as Bella walks. The distance between their houses is 2 miles, which is 10,560 feet, and Bella covers $2\frac{1}{2}$ feet with each step. How many steps will Bella 2 take by the time she meets Ella?

Problem 12.13
Jeremy's father drives him to school in rush hour traffic in 20 minutes. One day there is no traffic, so his father can drive him 18 miles per hour faster and gets him to school in 12 minutes. How far in miles is it to school?

Problem 12.14
Al walks down to the bottom of an escalator that is moving up and he counts 150 steps. His friend, Bob, walks up to the top of the escalator and counts 75 steps. If Al's speed of walking (in steps per unit time) is three times Bob's walking speed, how many steps are visible on the escalator at a given time? (Assume that this value is constant.)

Problem 12.15
Joe drives at a speed of 30 miles per hour for 60 miles. He also drives at 45 miles per hour for 15 miles. Find the average speed of his car.

Problem 12.16
Jones is chasing a car 800 meters ahead of him. He is on a horse moving at 50 km/h. If Jones catches up to the car in 4 minutes, how fast was the car moving?

Problem 12.17
David drives from his home to the airport to catch a flight. He drives 35 miles in the first hour, but realizes that he will be 1 hour late if he continues at this speed. He increases his speed by 15 miles per hour for the rest of the way to the airport and arrives 30 minutes early. How many miles is the airport from his home?

Problem 12.18
Jacob and Alexander are walking up an escalator in the airport. Jacob walks twice as fast as Alexander, who takes 18 steps to arrive at the top. Jacob, however, takes 27 steps to arrive at the top. How many of the upward moving escalator steps are visible at any point in time?

Problem 12.19
(AMC 10) Emily sees a ship traveling at a constant speed along a straight section of a river. She walks parallel to the riverbank at a uniform rate faster than the ship. She counts 210 equal steps walking from the back of the ship to the front. Walking in the opposite direction, she counts 42 steps of the same size from the front of the ship to the back. In terms of Emily's equal steps, what is the length of the ship?

Problem 12.20

(AIME) Points A, B, and C lie in that order along a straight path where the distance from A to C is 1800 meters. Ina runs twice as fast as Eve, and Paul runs twice as fast as Ina. The three runners start running at the same time with Ina starting at A and running toward C, Paul starting at B and running toward C, and Eve starting at C and running toward A. When Paul meets Eve, he turns around and runs toward A. Paul and Ina both arrive at B at the same time. Find the number of meters from A to B.

Problem 12.21

(AIME) Butch and Sundance need to get out of Dodge. To travel as quickly as possible, each alternates walking and riding their only horse, Sparky, as follows. Butch begins by walking while Sundance rides. When Sundance reaches the first of the hitching posts that are conveniently located at one-mile intervals along their route, he ties Sparky to the post and begins walking. When Butch reaches Sparky, he rides until he passes Sundance, then leaves Sparky at the next hitching post and resumes walking, and they continue in this manner. Sparky, Butch, and Sundance walk at 6, 4, and 2.5 miles per hour, respectively. The first time Butch and Sundance meet at a milepost, they are n miles from Dodge, and they have been traveling for t minutes. Find $n + t$.

§13 Sequence and Series

§13.1 Arithmetic Sequence

Definition 13.1. An **arithmetic sequence** is a sequence of numbers with the same difference between consecutive terms. $1, 4, 7, 10, 13, ..., 40$ is an arithmetic sequence because there is always a difference of 3 between consecutive terms.

Theorem 13.2

(n^{th} Term of a Sequence) Let a_n be the nth term of a sequence. Then

$$a_n = a_1 + (n-1)d$$

where d is the common difference and a_1 is the first term. From this, we can obtain the following formula for positive integers $n > m$

$$a_n = a_m + (n-m)d$$

Corollary 13.3

(Number of Terms in an Arithmetic Sequence) Let n be the number of terms in an arithmetic sequence $\{a_1, a_2, \ldots, a_n\}$. Then

$$n = \frac{a_n - a_1}{d} + 1$$

Corollary 13.4

(Average of Terms in an Arithmetic Sequence) Let x be the average of the terms in an arithmetic sequence $\{a_1, a_2, \ldots, a_n\}$.

$$x = \frac{a_1 + a_2 + \ldots + a_n}{n}$$

Then

$$x = \frac{a_1 + a_n}{2}$$

which essentially means

$$\text{Average of Terms} = \frac{\text{First Term} + \text{Last Term}}{2}$$

If the number of terms is even, $x =$ average of middle 2 terms.

If the number of terms is odd, $x =$ middle term.

Theorem 13.5

(Sum of All Terms in an Arithmetic Sequence)

$$S_n = \frac{a_1 + a_n}{2} \times n$$

which essentially means

Sum of All Terms = Average of Terms × Number of Terms

We can also substitute $a_n = a_1 + (n-1)d$ to get

$$S_n = \frac{2a_1 + (n-1)d}{2} \times n$$

Example 13.6

The sum of the first 5 terms of an arithmetic sequence is 65 and the sum of the first 10 terms of the same sequence is 255. Find the sum of the first 15 terms of the sequence.

Example 13.7

Find the value of $a_2 + a_4 + a_6 + \cdots + a_{98}$ if $a_1, a_2, a_3, \ldots$ is an arithmetic progression with common difference 1, and $a_1 + a_2 + a_3 + \cdots + a_{98} = 137$.

Theorem 13.8

Sum of Natural Numbers:

$$1 + 2 + 3 + \cdots + n = \frac{n(n+1)}{2}$$

Sum of Odd Numbers Formula

$$1 + 3 + 5 + \cdots + (2n-1) = n^2$$

Sum of Even Numbers Formula

$$2 + 4 + 6 + \cdots + 2n = n(n+1)$$

Sum of Squares Formula

$$1^2 + 2^2 + 3^2 + \cdots + n^2 = \frac{n(n+1)(2n+1)}{6}$$

Sum of Cubes Formula

$$1^3 + 2^3 + 3^3 + \cdots + n^3 = \left(\frac{n(n+1)}{2}\right)^2$$

§13.2 Geometric Sequence

A geometric sequence is a sequence of numbers with the same ratio between consecutive terms. For example,
$$1, 2, 4, 8, 16, 32, \ldots, 1024$$
is a geometric sequence because there is always a ratio of 2 between consecutive terms. In general, the terms of a geometric sequence $a_1, a_2, \ldots, a_n$ have the following properties. Let n represent the number of terms in the sequence and r represent the common ratio.

> **Theorem 13.9**
> (i^{th} Term in a Geometric Sequence)
> $$a_i = a_1 r^{i-1}, \quad 1 \leq i \leq n$$

> **Corollary 13.10**
> (Number of Terms in a Finite Geometric Sequence)
> $$n = \log_r \left(\frac{a_n}{a_1}\right) + 1$$

> **Corollary 13.11**
> (Sum of All Terms in a Finite Geometric Sequence)
> $$S_n = a_1 \times \frac{1 - r^n}{1 - r}$$
> If our sequence is infinite, our series converges to
> $$\frac{a_1}{1 - r}$$
> f and only if $|r| < 1$.

> **Example 13.12**
> (AMC 12) Call a 3-digit number geometric if it has 3 distinct digits which, when read from left to right, form a geometric sequence. Find the difference between the largest and smallest geometric numbers.

§13.3 Practice Problems

> **Problem 13.13**
> Suppose that $\{a_n\}$ is an arithmetic sequence with $a_1 + a_2 + \cdots + a_{100} = 100$ and $a_{101} + a_{102} + \cdots + a_{200} = 200$. What is the value of $a_2 - a_1$?

Problem 13.14
A sequence of three real numbers forms an arithmetic progression with a first term of 9. If 2 is added to the second term and 20 is added to the third term, the three resulting numbers form a geometric progression. What is the smallest possible value for the third term in the geometric progression?

Problem 13.15
The terms of an arithmetic sequence add to 715. The first term of the sequence is increased by 1, the second term is increased by 3, the third term is increased by 5, and in general, the kth term is increased by the kth odd positive integer. The terms of the new sequence add to 836. Find the sum of the first, last, and middle terms of the original sequence.

Problem 13.16
What is the 100th number in the arithmetic sequence $1, 5, 9, 13, 17, 21, 25, \ldots$?

Problem 13.17
The non-negative integers $a, b, c, d,$ and e form an arithmetic sequence. If their sum is 440, what is the largest possible value for e?

Problem 13.18
The sum of 25 consecutive even integers is $10,000$. What is the largest of these 25 consecutive integers?

Problem 13.19
A grocer makes a display of cans in which the top row has one can and each lower row has two more cans than the row above it. If the display contains 100 cans, how many rows does it contain?

§14 Mean, Median, and Mode

§14.1 Mean, Median, Mode

Definition 14.1. Suppose there is a sequence $\{a_n\}$ lined up in increasing order. Then we define the mean, mode, median, range as the following:

$$\text{Mean} = \text{Average of All Terms} = \frac{\text{Sum of All Terms}}{\text{Number of Terms}}$$

$$\text{Mode} = \text{Most Common Term(s)}$$

$$\text{Median} = \begin{cases} \text{middle number} & \text{if number of terms } = \text{odd} \\ \text{average of middle two numbers} & \text{if number of terms } = \text{even} \end{cases}$$

$$\text{Range} = \text{Largest Number} - \text{Smallest Number}$$

Example 14.2
Find the sum of the mean, median, mode, and range of $1, 9, 7, 1, 3, 5, 2$.

Solution. To solve this, we will apply the formulas above.

What is the first step we should do to make analyzing the numbers easier?

We should order them! Doing so, we end up with $1, 1, 2, 3, 5, 7, 9$.

What is the mean?

The mean is the sum of the numbers divided by the count of numbers. The sum of numbers is $1 + 1 + 2 + 3 + 5 + 7 + 9 = 28$. There are 7 numbers. Therefore, the mean is $\frac{28}{7} = 4$.

What is the median?

There are 7 numbers, so the median will just be the middle, or 4th number when the numbers are arranged in increasing order. The 4th number is 3, so the median is also 3. **What is the mode?** The only number that appears more than once is 1, so 1 is the mode. **What is the range?** The range will be the largest number, 9, minus the smallest number, 1, which is equal to $9 - 1 = 8$.

Therefore, the sum of the mean, median, mode, and range is $1 + 3 + 4 + 8 = 16$.

Example 14.3
(AIME) Let S be a list of positive integers—not necessarily distinct—in which the number 68 appears. The average (arithmetic mean) of the numbers in S is 56. However, if 68 is removed, the average of the remaining numbers drops to 55. What is the largest number that can appear in S?

---

§14.2 Mean, Median Mode Conditions Example

> **Example 14.4**
> A list of 6 distinct positive integers has a mean of 7, a median of 6, and a range of 9. Find the sum of the 2nd and 5th smallest numbers.

Solution. **What are the possibilities for the middle 2 numbers, or the 3rd and 4th smallest numbers?**

The average of these 2 numbers must be 6, or the sum must be $2 \times 6 = 12$.

What must be the sum of the remaining 4 numbers?

The mean of all 6 numbers is 7, so their sum is $6 \times 7 = 42$. The sum of the middle 2 numbers is 12, so the sum of the remaining 4 numbers must be $42 - 12 = 30$. Next, let's consider the cases for the 2 middle numbers that result in a sum of 12. The following possibilities exist:

Case 1: 3rd smallest number: 6, 4th smallest number: 6

This is not possible because the numbers are distinct.

Case 2: 3rd smallest number: 4, 4th smallest number: 8

In this case, there are very few possibilities for the lowest 2 numbers. Since they must be positive, less than 4, and distinct, so only possibilities are $(1,2), (1,3)$, and $(2,3)$. In addition, the range must also be 9. Therefore, the largest number will be the lowest number plus the range, which can be $1 + 9 = 10$ or $2 + 9 = 11$. We can consider each of the cases for the largest and smallest numbers, however, we should first check if this case is even possible.

Is it possible for the sum of the $1^{st}, 2^{nd}, 5^{th}$, and 6th smallest numbers to be 30 in this case?

We can consider each of the cases for the largest and smallest numbers, however, we should first check if having a mean of 7 is even possible. The largest number can be at most 11, and the 2nd largest number can be at most 10 since it can't be the same as the largest number. This means the sum of these 4 numbers can be at most $2 + 3 + 10 + 11 = 26$, which isn't large enough to be 30. Therefore, this case doesn't work.

Case 3: 3rd smallest number 3, 4th smallest number 9

The smallest 2 numbers must be 1 and 2, since all of the numbers must be positive and less than 3. Then, by similar logic to the previous case, the largest value is $1 + 9 = 10$. The 2nd largest value can be at most 9. This means the sum is at most $1 + 2 + 9 + 10 = 22$, which is again, not large enough.

Case 4: 3rd smallest number 2 or 1, 4th smallest number 10 or 11, respectively

For the other cases, the 3rd smallest number will be less than 3. However, we must have 2 distinct positive values less than the 3rd smallest number (the 1st and 2nd smallest numbers). We can see that if the 3rd smallest value is less than 3 and is 1 or 2, then there are 0 or 1 distinct positive values less than this number. Therefore, these cases don't work.

Case 5: 3rd smallest number: 5, 4th smallest number 7

Is it possible for the smallest number to be 2? Let's see if it's possible for the sum of smallest 2 numbers and largest 2 numbers to be 30 in this case. The largest possible value for the 2nd smallest number is 4, since it has to be less than 5. The largest number must be the smallest number plus the range, which is $2 + 9 = 11$, and the 2nd largest number can be at most 10. Therefore, sum of smallest 2 numbers and largest 2 numbers is at most $2 + 4 + 10 + 11 = 27$. Therefore, the smallest number cannot be 2.

If the smallest number is 1, the sum will be even lower since the largest will be 10, and the 2nd largest number will be at most 9. This would result in a sum even lower than in the previous possibility. Therefore, the smallest number must be 3, and since the 2nd smallest number has to be greater than the smallest number (3) and less than the 3rd smallest number (5), it must be 4. In addition, the largest number will be $3 + 9 = 12$. Then, let the 2nd largest number be x.

What value of x will result in the sum smallest 2 numbers and largest 2 numbers being 30?

Adding all of these numbers, we get that this sum is $3 + 4 + 12 + x = 30$. This means that the 2nd largest number is 11. This case works because this number is smaller than the largest number (12), and larger than the 4th smallest number (7). The 6 numbers (in increasing order) are $3, 4, 5, 7, 9, 11, 12$ satisfy all 3 conditions. The sum of the 2nd and 5th smallest numbers are $4 + 9 = 13$.

Example 14.5

(AMC 10) The mean, median, unique mode, and range of a collection of eight integers are all equal to 8. The largest integer that can be an element of this collection is what?

---

§14.3 Practice Problems

Problem 14.6

Four students take an exam. Three of their scores are $70, 80,$ and 90. If the average of their four scores is 70, then what is the remaining score?

Problem 14.7

The following bar graph represents the length (in letters) of the names of 19 people. What is the median length of these names?

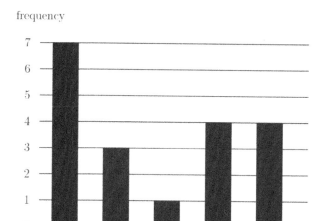

Problem 14.8
Mr. Garcia asked the members of his health class how many days last week they exercised for at least 30 minutes. The results are summarized in the following bar graph, where the heights of the bars represent the number of students.

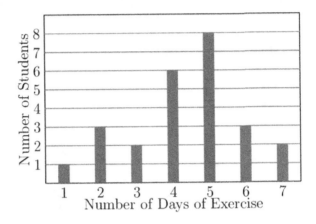

What was the mean number of days of exercise last week, rounded to the nearest hundredth, reported by the students in Mr. Garcia's class?

Problem 14.9
(AMC 8) Hammie is in the grade and weighs 106 pounds. Her quadruplet sisters are tiny babies and weigh 5, 5, 6, and 8 pounds. Which is greater, the average (mean) weight of these five children or the median weight, and by how many pounds?

Problem 14.10
(AMC 8) What is the sum of the mean, median, and mode of the numbers 2, 3, 0, 3, 1, 4, 0, 3?

Problem 14.11
(AMC 8) The mean, median, and unique mode of the positive integers 3, 4, 5, 6, 6, 7, and x are all equal. What is the value of x?

Problem 14.12
(AMC 8) The mean of a set of five different positive integers is 15. The median is 18. What is the maximum possible value of the largest of these five integers?

Problem 14.13
(AMC 8) One day the Beverage Barn sold 252 cans of soda to 100 customers, and every customer bought at least one can of soda. What is the maximum possible median number of cans of soda bought per customer on that day?

Problem 14.14
(AMC 8) The harmonic mean of a set of non-zero numbers is the reciprocal of the average of the reciprocals of the numbers. What is the harmonic mean of $1, 2$, and 4?

Problem 14.15
(AMC 8) How many subsets of two elements can be removed from the set $\{1, 2, 3, 4, 5, 6, 7, 8, 9, 10, 11\}$ so that the mean (average) of the remaining numbers is 6?

Problem 14.16
(AMC 10) When the mean, median, and mode of the list

$$10, 2, 5, 2, 4, 2, x$$

are arranged in increasing order, they form a non-constant arithmetic progression. What is the sum of all possible real values of x?

Problem 14.17
Suppose that S is a finite set of positive integers. If the greatest integer in S is removed from S, then the average value (arithmetic mean) of the integers remaining is 32. If the least integer in S is also removed, then the average value of the integers remaining is 35. If the greatest integer is then returned to the set, the average value of the integers rises to 40. The greatest integer in the original set S is 72 greater than the least integer in S. What is the average value of all the integers in the set S?

Problem 14.18
The mean of the numbers $2, 0, 1, 5$, and x is an integer. Find the smallest possible positive integer value for x.

Problem 14.19
(AMC 10) The mean, median, and mode of the 7 data values $60, 100, x, 40, 50, 200, 90$ are all equal to x. What is the value of x?

Problem 14.20

(AMC 10) When 15 is appended to a list of integers, the mean is increased by 2. When 1 is appended to the enlarged list, the mean of the enlarged list is decreased by 1. How many integers were in the original list?

Problem 14.21

The mean of a list of nine numbers is 17, and the modes are a, b and c. If $a+4, 1+b$ and $c-8$ are distinct numbers in the list, and none of them are modes of the list, then what is the value of $3(a+b+c)$?

Problem 14.22

6 positive integers have a median of 5.5, two distinct modes, a range of 5, and a mean of $\frac{35}{6}$. Find the sum of all possible values for the 2nd largest integer.

§15 Telescoping

§15.1 Telescoping

Expand the first few and last few terms, and cancel out any terms you see.

Generally, whenever you have long expressions that seem to be hard or impossible to compute manually, telescoping is probably at play.

§15.2 Telescoping Sum and Product

Example 15.1
Evaluate the product
$$\frac{1}{2} \times \frac{2}{3} \times \frac{3}{4} \times \cdots \times \frac{19}{20}.$$

Example 15.2
Find the value of the product
$$\left(1 + \frac{1}{1}\right)\left(1 + \frac{1}{2}\right)\left(1 + \frac{1}{3}\right) \cdots \left(1 + \frac{1}{9}\right)\left(1 + \frac{1}{10}\right)$$

Example 15.3
Determine the value of
$$\left(1 - \frac{1}{1000}\right)\left(1 - \frac{1}{999}\right)\left(1 - \frac{1}{998}\right) \cdots \left(1 - \frac{1}{3}\right)\left(1 - \frac{1}{2}\right)$$

Example 15.4
Find the sum
$$\left(\frac{1}{2} - \frac{1}{3}\right) + \left(\frac{1}{3} - \frac{1}{4}\right) + \cdots + \left(\frac{1}{18} - \frac{1}{19}\right) + \left(\frac{1}{19} - \frac{1}{20}\right)$$

Example 15.5
Determine the value of
$$\frac{1}{1 \times 2} + \frac{1}{2 \times 3} + \frac{1}{3 \times 4} + \cdots + \frac{1}{98 \times 99} + \frac{1}{99 \times 100}$$

Example 15.6
Evaluate
$$\frac{1}{3} + \frac{1}{15} + \frac{1}{35} + \frac{1}{63} + \frac{1}{99} + \frac{1}{143}$$

Partial Fraction

§15.3 Practice Problems

Problem 15.7
What is the value of $4(-1 + 2 - 3 + 4 - 5 + 6 - 7 + \cdots + 1000)$?

Problem 15.8
Find the value of the expression $100 - 98 + 96 - 94 + 92 - 90 + \cdots + 8 - 6 + 4 - 2$.

Problem 15.9
What is the value of $1 + 3 + 5 + \cdots + 2017 + 2019 - 2 - 4 - 6 - \cdots - 2016 - 2018$?

Number Theory

§16 Primes and Divisibility

§16.1 Primes

Definition 16.1. Primes are numbers that have exactly two factors: 1 and the number itself. For example, 2, 3, 5, 7, 11, 13, 17, 19, 23 are all primes

Note: 1 is not a prime and 2 is the only even prime.

§16.2 Divisibility Rules

Divisibility rule for 2 :: Last digit is even

Divisibility rule for 3 :: Sum of digits is divisible by 3

Divisibility rule for 4 : The number formed by the last 2 digits is divisible by 4

Divisibility rule for 5 : Last digit is 0 or 5

Divisibility rule for 6 : The number is divisible by both 2 and 3

Divisibility rule for 7 : Take out factors of 7 until you reach a small number that is either divisible or not divisible by 7

Example: 2240

We subtract 2100, which is divisible by 7 (7×300) to get $2240 - 2100 = 140$. Then, we see that 140 is also divisible by 7 as it is 7×20. Therefore, 2240 is also divisible by 7.

Example: 28146

We subtract 28000, which is divisible by 7 (7×4000) to get $28146 - 28000 = 146$. Then, we subtract 140 as it is also divisible by 7 to get 6. 6 is clearly not divisible by 7, so 28146 is also not divisible by 7.

Divisibility rule for 8 : The number formed by the last 3 digits is divisible by 8

Divisibility rule for 9 : Sum of digits is divisible by 9

Divisibility rule for 10 : Last digit is 0

Divisibility rule for 11 : Calculate the sum of odd positioned digits (O) and even positioned digits (E). If $O - E$ is divisible by 11, then the number is also divisible by 11. Don't forget that $O - E$ can be negative.

Example: 1331

First, we calculate the sum of the odd positioned digits, or the sum of the 1st and 3rd digits from the left. We find $O = 1 + 3 = 4$. Next, we calculate the sum of the even positioned digits, or the sum of the 2nd and 4th digits from the left. We find $E = 3 + 1 = 4$. The difference of numbers, $O - E = 4 - 4 = 0$, and 0 is a multiple of 11, so 1331 is divisible by 11.

Example: 629321

First, we calculate the sum of the odd positioned digits, or the sum of the $1^{\text{st}}, 3^{\text{rd}}$, and 5th digits from the left. We find $O = 6 + 9 + 2 = 17$. Next, we calculate the sum of the even positioned digits, or the sum of the $2^{\text{nd}}, 4^{\text{th}}$, and 6th digits. We find $E = 2 + 3 + 1 = 6$. The difference $O - E = 17 - 6 = 11$, which is clearly a multiple of 11, so 629321 is divisible by 11.

Divisibility rule for 12: divisible by 3 and 4

Divisibility rule for 15: Divisible by 3 and 5

§16.3 Primes Factorization

Prime factorization is a way to express each number as a product of primes.

> **Example 16.2**
> Find the prime factorization of 117.

> **Example 16.3**
> (AMC 8) Let Z be a 6-digit positive integer, such as 247247, whose first three digits are the same as its last three digits taken in the same order. Which of the following numbers must also be a factor of Z?

> **Example 16.4**
> (MATHCOUNTS) What four-digit number has tens digit 2 and units digit 8, is a multiple of 16, and when its digits are reversed the result is also a multiple of 16?

> **Example 16.5**
> (AMC 8) A number is called *flippy* if its digits alternate between two distinct digits. For example, 2020 and 37373 are *flippy*, but 3883 and 123123 are not. How many five-digit flippy numbers are divisible by 15?

> **Example 16.6**
> The 3 digit number $\overline{A3B}$ is divisible by 15. What is the sum of all such possible numbers?

We can use our knowledge of divisibility rules to solve this.

For a number to be divisible by 15, what other numbers must it be divisible by?

We can see that $15 = 3 \times 5$, so for a number to be divisible by 15, it must also be divisible by 3 and 5. In order to solve this problem, we will look at the divisibility rule for 3 and 5.

Which divisibility rule should we look at first?

If we look at the divisibility rule for 3 first, we will end with many possibilities for A and B.

On the other hand, what does the divisibility rule for 5 tell us?

The divisibility rule for 5 tells us that B must be 0 or 5. This is much easier to use than the divisibility rule for 3 as now, we only have 2 possibilities to consider. Because we have already checked for the divisibility of 5, we must now check if the number is divisible by 3.

Case 1: $B = 0$

If $B = 0$, then the number will be $\overline{A30}$. For this number to be divisible by 3, $A + 3 + 0 = A + 3$ must be divisible by 3. Because 3 is a multiple of 3, A must also be a multiple of 3. Therefore, A has 3 possibilities and can be 3, 6, or 9. The digit A can't be 0 since our number has to be a 3 digit number. This gives the following possible numbers: $330, 630, 930$.

Case 2: B = 5

If $B = 5$, then the number will be $\overline{A35}$. For this number to be divisible by 3, $A + 3 + 5 = A + 8$ must be divisible by 3. Because 8 leaves a remainder of 2 when divided by 3, A must leave a remainder of 1 when divided by 3 for the sum to be a multiple of 3. Therefore, A has 3 possibilities and can be 1, 4, or 7. This gives the following possible numbers: $135, 435, 735$. Summing all numbers, we get $330 + 630 + 930 + 135 + 435 + 735 = 3195$

Example 16.7
The number $\overline{3AB76}$ is divisible by 264, where A and B are digits. Find $A + B$.

Solution. **For a number to be divisible by 264, what other numbers must it be divisible by?**

To figure this out, we first find the prime factorization of 264. First, we divide out by powers of 2. Doing this repeatedly we get $264 = 2 \times 132 = 2^2 2 \times 66 = 2^3 \times 33$ Then, we can notice that the sum of digits of 33 is $3 + 3 = 6$, which is a multiple of 3, so 33 is a multiple of 3. Dividing by 3, we are then left with 11, which is clearly a prime number. Therefore, the prime factorization is $2^3 \times 3 \times 11$. Therefore, for the number to be divisible by 264, it must be divisible by $2^3 = 8, 3$, and 11.

First, what information does $\overline{3AB76}$ being divisible by 8 give us?

By the divisibility rule for 8, the last 3 digits, or $\overline{B76}$, must be divisible by 8. Notice that 76 leaves a remainder of 4 when divided by 8. Therefore, $B = 0$ doesn't work. When B is 1, then we are adding 100 to the last 3 digits. 100 also leaves a remainder of 4 when divided by 8. So $176 = 76 + 100$ will leave a remainder of $4 + 4 = 8$ or 0 when divided by 8, so 176 is divisible by 8.

Next, we can see that when B is 2, our number will be 276, which is equal to $100 + 176$. Since 176 is divisible by 8 and 100 leaves a remainder of 4 when divided by 8, 276 will leave a remainder of $0 + 4 = 4$ when divided by 8.

Do you notice a pattern with the remainders here?

Whenever B is even, it appears that $\overline{B76}$ leaves a remainder of 4 when divided by 8, and whenever B is odd, $\overline{B76}$ is divisible by 8. This is because 176 is divisible by 8, adding 100 will make the remainder

4 when divided by 8. Therefore, by adding 200, the remainder will become $4 + 4 = 8$ or 0. Therefore, only odd values of B work.

Next, what does $\overline{3AB76}$ being divisible by 11 tell us?

Using the divisibility rule for 11, we have that the sum of the odd positioned digits, $O = 3 + B + 6 = 9 + B$ and the sum of the even positioned digits, $E = A + 7$. Therefore, $O - E = 2 + B - A$ must be divisible by 11.

What are the possible values for $O - E = 2 + B - A$?

Let's suppose it is equal to -11. Then $2 + B - A = -11$ which means $A - B = 13$. Since A and B are digits from 0 to 9, this is clearly impossible. Any other negative multiple of 11 such as $-22, -33$, etc. will also mean that $A - B$ equals something larger than 9. Next, let's suppose it is equal to 0. Then $2 + B - A = 0$ which means $A - B = 2$. Since we have that B is odd from it being a multiple of 8, the only possibilities are:

1. $A = 3, B = 1$
2. $A = 5, B = 3$
3. $A = 7, B = 5$
4. $A = 9, B = 7$

If $B = 9$, then $A = 11$, which is not possible as A needs to be 1 digit. Finally, let's suppose $O - E$ equals 11. Then $2 + B - A = 11$ which means $B - A = 9$. Because A and B are digits from 0 to 9, the only possibility is when $A = 0$ and $B = 9$. Any other positive multiple of 11 such as 22, 33, etc. will mean that $B - A$ equals something larger than 9.

How do we use the fact that $\overline{3AB76}$ is a multiple of 3 to finish the problem?

Notice how we only have 5 cases, so we can easily manually check if each one is a multiple of 3 by using its divisibility rule.

Case 1: $A = 3, B = 1$

For this case, the number is 33176. The sum of digits is $3 + 3 + 1 + 7 + 6 = 20$, which is not a multiple of 3, so this case does not work.

Case 2: $A = 5, B = 3$

For this case, the number is 35376. The sum of digits is $3 + 5 + 3 + 7 + 6 = 24$, which is a multiple of 3, so this case does work.

Case 3: $A = 7, B = 5$

For this case, the number is 37576. The sum of digits is $3 + 7 + 5 + 7 + 6 = 28$, which is not a multiple of 3, so this case does not work.

Case 4: $A = 9, B = 7$

For this case, the number is 39776. The sum of digits is $3 + 9 + 7 + 7 + 6 = 32$, which is not a multiple of 3, so this case does not work.

Case 5: $A = 0, B = 9$

For this case, the number is 30976. The sum of digits is $3+0+9+7+6 = 25$, which is not a multiple of 3, so this case does not work. Therefore, $A = 5$ and $B = 3$ is the only case that works, so our answer is $5 + 3 = 8$.

§16.4 Practice Problems

Problem 16.8
Which of the following numbers has the smallest prime factor?

(A) 55   (B) 57   (C) 58   (D) 59   (E) 61

Problem 16.9
What is the sum of the two smallest prime factors of 250?

Problem 16.10
How many three-digit numbers are divisible by 13?

Problem 16.11
The sum of two prime numbers is 85. What is the product of these two prime numbers?

Problem 16.12
The 7-digit numbers $\overline{74A52B1}$ and $\overline{326AB4C}$ are each multiple of 3. Which of the following could be the value of C?

(A) 1   (B) 2   (C) 3   (D) 5   (E) 8

Problem 16.13
The 5-digit number $\overline{2018U}$ is divisible by 9. What is the remainder when this number is divided by 8?

Problem 16.14
(AMC 8) What is the smallest positive integer that is neither prime nor square and that has no prime factor less than 50?

Problem 16.15
(AMC 8) For any positive integer M, the notation $M!$ denotes the product of the integers 1 through M. What is the largest integer n for which 5^n is a factor of the sum $98! + 99! + 100!$?

Problem 16.16
(AMC 8) Three members of the Euclid Middle School girls' softball team had the following conversation.

- Ashley: I just realized that our uniform numbers are all 2-digit primes.
- Brittany: And the sum of your two uniform numbers is the date of my birthday earlier this month.
- Caitlin: That's funny. The sum of your two uniform numbers is the date of my birthday later this month.
- Ashley: And the sum of your two uniform numbers is today's date.

What number does Caitlin wear?

Problem 16.17
The number $\overline{AB962C}$ is divisible by 792 where A, B, C are digits. What is the value of A?

Problem 16.18
(AMC 8) The digits $1, 2, 3, 4$, and 5 are each used once to write a five-digit number $\overline{PQRST}$. The three-digit number $\overline{PQR}$ is divisible by 4, the three-digit number $\overline{QRS}$ is divisible by 5, and the three-digit number $\overline{RST}$ is divisible by 3. What is P?

Problem 16.19
(AMC 10) The base-ten representation for $19!$ is $121,6T5,100,40M,832,H00$, where T, M, and H denote digits that are not given. What is $T + M + H$?

Problem 16.20
(AMC 10) Call a positive integer an uphill integer if every digit is strictly greater than the previous digit. For example, 1357,89, and 5 are all uphill integers, but 32, 1240, and 466 are not. How many uphill integers are divisible by 15?

Problem 16.21
What is the prime factorization of 420?

Problem 16.22
What is the prime factorization of 1001?

Problem 16.23
(AMC 8) The number 64 has the property that it is divisible by its units digit. How many whole numbers between 10 and 50 have this property?

Problem 16.24
Ayasha, Beshkno, and Chenoa were all born after 2000. Each of them was born in a year after 2000 that is divisible by exactly one of the prime numbers 2, 3 or 5. Each of these primes is a divisor of one of the birth years. What is the least possible sum of their birth years?

Essential Academy (June 2023) — Introduction to AMC 8

§17 Factors

§17.1 Number of Factors

Theorem 17.1
(Number of Factors of a Number) Let n be a positive integer and $p_1^{e_1} p_2^{e_2} \cdots p_k^{e_k}$ be its prime factorization. Then the number of factors or divisors of n is given by the formula
$$(e_1 + 1)(e_2 + 1) \cdots (e_k + 1)$$

Example 17.2
How many factors does the number 112 have?

Example 17.3
How many positive factors does the number 144 have that are perfect squares?

Example 17.4
(AMC 8) How many positive integer factors of 2020 have more than 3 factors? (As an example, 12 has 6 factors, namely $1, 2, 3, 4, 6,$ and 12.)

§17.2 Sum of Factors

Theorem 17.5
(Sum of the Factors of a Number) Let n be a positive integer and $p_1^{e_1} p_2^{e_2} \cdots p_k^{e_k}$ be its prime factorization. Then the sum of the factors or divisors of n is given by the formula
$$(1 + p_1 + \cdots + p_1^{e_1})(1 + p_2 + \cdots + p_2^{e_2}) \cdots (1 + p_k + \cdots + p_k^{e_k})$$

Example 17.6
What is the sum of factors of 112?

§17.3 Product of Factors

Theorem 17.7
(Product of the Factors of a Number) Let n be a positive integer and let $\tau(n)$ be the number of factors of n. Then the product of the factors of n is given by the formula
$$\sqrt{n^{\tau(n)}} = n^{\frac{\tau(n)}{2}}$$

Example 17.8
Find the product of the factors of 20.

Example 17.9
Let a be the product of all odd factors of 54. Let b be the sum of all even factors of 36. Find $a - b$.

§17.4 Practice Problems

Problem 17.10
(AMC 8) What is the sum of the prime factors of 2010?

Problem 17.11
(AMC 8) How many positive factors does $23,232$ have?

Problem 17.12
(AMC 8) For any positive integer M, the notation $M!$ denotes the product of the integers 1 through M. What is the largest integer n for which 5^n is a factor of the sum $98! + 99! + 100!$?

(A) 23 (B) 24 (C) 25 (D) 26 (E) 27

Problem 17.13
(AMC 8) On June 1, a group of students is standing in rows, with 15 students in each row. On June 2, the same group is standing with all of the students in one long row. On June 3, the same group is standing with just one student in each row. On June 4, the same group is standing with 6 students in each row. This process continues through June 12 with a different number of students per row each day. However, on June 13, they cannot find a new way of organizing the students. What is the smallest possible number of students in the group?'

Problem 17.14
How many positive multiples of 1001 have less than 15 factors?

Problem 17.15
(AMC 10) Let $N = 34 \times 34 \times 63 \times 270$. What is the ratio of the sum of the odd divisors of N to the sum of the even divisors of N?

Problem 17.16
How many positive factors does the number 60 have?

Problem 17.17
Find the sum of factors of 320.

§18 GCF and LCM

§18.1 GCF and LCM Fundamentals

The Greatest Common Divisor (GCD) of two or more integers (which are not all zero) is the largest positive integer that divides each of the integers. Note: This is also known as GCF (Greatest Common Factor), and the terms GCF and GCD are often used interchangeably.

The Least Common Multiple (LCM) of two or more integers (which are not all zero) is the smallest positive integer that is divisible by both the numbers.

> **Example 18.1**
> (AMC 8) What is the ratio of the least common multiple of 180 and 594 to the greatest common factor of 180 and 594?

§18.2 GCF and LCM Product

> **Theorem 18.2**
> The product of GCD and LCM of two numbers is equal to the product of the two numbers:
> $$\gcd(m, n) \cdot \text{lcm}(m, n) = mn$$

> **Example 18.3**
> Suppose we have 2 numbers m and n such that $mn = 1260$ and $\text{lcm}(m, n) = 210$. Find $\gcd(m, n)$.

§18.3 GCF and LCM Properties

> **Theorem 18.4**
> Properties:
>
> 1. $\gcd(ac, bc) = c \cdot \gcd(a, b)$
>
> 2. $\text{lcm}(ac, bc) = c \cdot \text{lcm}(a, b)$
>
> 3. If a number is divisible by two numbers a and b, it will also be divisible by $lcm(a, b)$
>
> 4. If a number divides two numbers a and b, it will also divide $\gcd(a, b)$.

> **Example 18.5**
> Let a be the smallest positive perfect cube that's divisible by $15, 22,$ and 176. If $a = b^3$, find b.

§18.4 Euclidean Algorithm

> **Theorem 18.6**
>
> (Euclidean Algorithm) The Euclidean Algorithm states that
> $$\gcd(x, y) = \gcd(x - ky, y)$$
> for any positive integers $x > y$ and positive integer k.

> **Example 18.7**
>
> Find $\gcd(186, 92)$.

Solution. We can apply the Euclidean Algorithm multiple times to easily find the GCD of large numbers since after applying the Euclidean algorithm, we will have two smaller numbers, and we can repeatedly apply the Euclidean Algorithm again until we get two very small numbers.

$$\begin{aligned}\gcd(186, 92) &= \gcd(186 - (2 \cdot 92), 92) \\ &= \gcd(2, 92) \\ &= \gcd(2, (92 - (2 \cdot 46))) \\ &= \gcd(2, 0) \\ &= 2\end{aligned}$$

> **Example 18.8**
>
> 3 numbers a, b, and c satisfy the conditions that $\gcd(a, b) = 4$, $\gcd(b, c) = 18$, and $\text{lcm}(a, c) = 144$. Find the value of $a + c$.

Solution. **What must be true about a and b if $\gcd(a, b) = 4$?**

If $\gcd(a, b) = 4$, then both a and b must be multiples of 4. Similarly, if $\gcd(b, c) = 18$, then both b and c must be multiples of 18.

What do both of these conditions combined tell us about b?

If b is a multiple of 4 and a multiple of 18, it also must be a multiple of the least common multiple of 4 and 18 by the above theorem. This means b is a multiple of 36. The reason this is true is because to be a multiple of 18, it must also be as multiple of 9 and 2. However, a number that is a multiple of 4 is already a multiple of 2, so therefore it just has to be a multiple of $9 \times 4 = 36$.

We know that a must be a multiple of 4. What happens if a is also a multiple of 3, or even 9?

Keep in mind that $\gcd(a, b) = 4$ means that not only both numbers share a factor of 4, but that no other factors are shared between the numbers. From earlier, we know that b is a multiple of 36. Therefore, if a has any factors of 3, then the gcd will also have factors of 3 since both a and b would then have factors of 3. Therefore, a cannot have any factors of 3.

Similarly, what must be true about c?

We know c is a multiple of 18, which means it's a multiple of 9 and a multiple of 2. On the other hand, b is a multiple of 36. The gcd is 18, so if c is a multiple of 36, then the gcd would have been 36, which is not possible. Therefore, c cannot be a multiple of 36. We already know c is a multiple of 9, that means c is not a multiple of 4.

Finally, how do we use the condition that $\text{lcm}(a, c) = 144$?

First, let's find the prime factorization of 144. First, we divide by factors of 2 to get
$$144 = 2 \cdot 72 = 2^2 \cdot 36 = 2^3 \cdot 18 = 2^4 \cdot 9$$
9 is 3^2, so prime factorization is $2^4 \cdot 3^2$.

For $\text{lcm}(a, c)$ to have a factor of 2^4, how many factors of 2 must a and c each have?

Remember from earlier that c is a multiple of 2, but not a multiple of 4. Therefore, c has 1 factor of 2. However, for the lcm to have a factor of 2^4, one of the numbers a or c must have exactly 4 powers of 2 in the prime factorization since the lcm is found by taking the maximum powers of each prime in the prime factorization of both numbers. Therefore, a must have exactly 4 powers of 2. It can't be more than 4 powers of 2 because then the lcm would also have more than 4 powers of 2. Also, c is a multiple of 2, but not 4, it will have exactly 1 power of 2 in it's prime factorization.

Next, for $\text{lcm}(a, c)$ to have a factor of 3^2, how many factors of 3 must a and c each have?

Earlier, we found that a has no factors of 3 and c must be a multiple of 9. Therefore, c must have exactly 2 powers of 3 in it's prime factorization. Again, it can't have any more than 2 powers of 3 since that would result in lcm also having more than 2 powers of 3.

Finally, can either a or c have any other prime factors besides 2 or 3?

If a or c had a factor of 5, 7, or some other prime, then the lcm would also have that prime in it's prime factorization. But earlier, we found that the prime factorization $2^4 \cdot 3^2$ only contains factors of 2 and 3. Therefore, this is impossible.

What are the values of a and c that work?

Since a has 4 powers of 2, no powers of 3, and no powers of any other prime, the prime factorization is 2^4. Therefore, $a = 16$.

Similarly, since b has 1 power of 2, 2 powers of 3, and no powers of any other prime, the prime factorization is $2 \cdot 3^2$. Therefore, $b = 18$.

The sum is therefore $16 + 18 = 34$.

§18.5 Practice Problems

Problem 18.9
(AMC 8) The least common multiple of a and b is 12, and the least common multiple of b and c is 15. What is the least possible value of the least common multiple of a and c?

Problem 18.10
What is the smallest positive integer that is divisible by 2 and 3 that consists entirely of 2's and 3's, with at least one of each?

Problem 18.11
What is the smallest five-digit integer divisible both by 8 and by 9?

Problem 18.12
Which of the following is/are divisible by 11?

- 3,951
- 907,654
- 14,256

Problem 18.13
What is the largest five-digit multiple of 11?

Problem 18.14
Find the remainder when $1,234,567$ is divided by 11?

Problem 18.15
Find the smallest positive integer greater than $90,000$ that is divisible by 11.

§19 Modular Arithmetic

§19.1 Modular Definition

Definition 19.1.
$$a \equiv b \pmod{n}$$
means the number a leaves the same remainder as b when divided by n

§19.2 Product Rule

> **Theorem 19.2**
> If $a \equiv x \pmod{n}$ and $b \equiv y \pmod{n}$, then $ab \equiv xy \pmod{n}$.

> **Example 19.3**
> (AMC 8) If n is an even positive integer, the *double factorial* notation $n!!$ represents the product of all the even integers from 2 to n. For example, $8!! = 2 \cdot 4 \cdot 6 \cdot 8$. What is the units digit of the following sum?
> $$2!! + 4!! + 6!! + \cdots + 2018!! + 2020!! + 2022!!$$
> **(A)** 0   **(B)** 2   **(C)** 4   **(D)** 6   **(E)** 8

§19.3 Exponent Rule

> **Theorem 19.4**
> If $a \equiv b \pmod{n}$ then
> $$a^k \equiv b^k \pmod{n}$$
> for any positive integer k.

> **Example 19.5**
> (AMC 8) When 1999^{2000} is divided by 5, what is the remainder?

Solution. **What mod value is equivalent to** $1999 \pmod 5$?

1999 leaves a remainder of 4 when divided by 5 so we could say that $1999 \equiv 4 \pmod 5$. However, then we would still have to calculate the remainder when 4^{2000} is divided by 5.

Instead, what negative mod is equivalent to $1999 \pmod 5$?

Since 1999 is 1 less than 2000, a multiple of 5, we can say that $1999 \equiv -1 \pmod 5$. Then, we have to find the remainder when $(-1)^{2000}$ is divided by 5, which is
$$((-1)^2)^{1000} \equiv 1^{1000} \equiv 1 \pmod 5$$
so it leaves a remainder of 1 when divided by 5.

§19.4 Multiple Modular Congruence

Example 19.6
Find the largest possible number less than 100 that leaves a remainder of 1 when divided by 5 and a remainder of 1 when divided by 7.

Solution. **If $x \equiv 1 \pmod 5$ and $x \equiv 1 \pmod 7$, then what is the value of $x - 1 \pmod 5$ and $\pmod 7$?**

Subtracting 1 from both sides of the equation $x \equiv 1 \pmod 5$ gives us $x - 1 \equiv 0 \pmod 5$. Similarly, we get $x - 1 \equiv 0 \pmod 7$. Therefore, $x - 1$ is a multiple of both 5 and 7.

If $x - 1$ is a multiple of both 5 and 7, what else must $x - 1$ be a multiple of?

If $x - 1$ is a multiple of both 5 and 7, $x - 1$ will also be a multiple of $5 \cdot 7 = 35$. Therefore, we have $x - 1 \equiv 0 \pmod{35}$ or $x \equiv 1 \pmod{35}$. From here, we can see that the possible values less than 100 are $1, 36, 71, 106$, etc. Clearly, 71 is the largest possible value less than 100.

§19.5 Digit Cycle

Example 19.7
Find the units digit of 2^{1026}.

§19.6 Practice Problems

Problem 19.8
What is the units digit of 13^{2012}?

Problem 19.9
What is the units digit of $19^{19} + 99^{99}$?

Problem 19.10
(AMC 8) What is the tens digit of 7^{2011}?

Problem 19.11
What is the smallest number greater than 2 that leaves a remainder of 2 when divided by $3, 4, 5$, or 6?

Problem 19.12
(AMC 8) The number N is a two-digit number. When N is divided by 9, the remainder is 1. When N is divided by 10, the remainder is 3. What is the remainder when N is divided by 11?

Problem 19.13

(AMC 8) The product of the two 99-digit numbers

$$303,030,303,...,030,303 \text{ and } 505,050,505,...,050,505$$

has thousands digit A and units digit B. What is the sum of A and B?

Problem 19.14

(AMC 8) How many positive three-digit integers have a remainder of 2 when divided by 6, a remainder of 5 when divided by 9, and a remainder of 7 when divided by 11?

Geometry

§20 Angle Chasing

§20.1 Angle Chasing Tricks

> **Theorem 20.1**
>
> (Angle Chasing Tricks) Here are some nice angle chasing ideas used to solve problems:
>
> - Sum of Angles in Triangle is 180
>
> - A triangle with 2 angles equal will have their corresponding sides equal and a triangle with 2 sides equal will have their corresponding angles equal (isosceles triangle)
>
> - Opposite angles in intersecting lines are equal
>
> - Corresponding angles in parallel lines are equal
>
> - The angle made by the arc at the center of the circle is double the angle made by the arc at the boundary of the circle.

Angles:

Complementary Angle

Supplementary Angle

Intersecting Line

Parallel Line

Corresponding Angle

Vertical Angle

Alternative Interior Angle

Alternative Exterior Angle

Example 20.2
The degree measure of angle A is

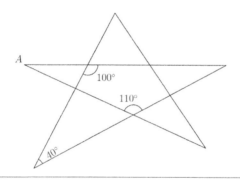

Example 20.3
In acute $\triangle ABC$, points D, E are inside triangle ABC such that $DE \parallel BC$ with B closer to point D than to point E. $\angle AED = 80°$, $\angle ABD = 10°$, and $\angle CBD = 40°$. Find the measure of $\angle BAE$, in degrees.

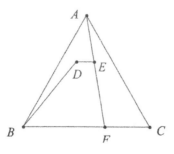

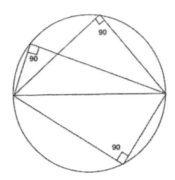

§20.2 Inscribed Angle

Theorem 20.4

(Inscribed Arc Theorem) The angle formed by an arc in the center or the arc angle is double of the angle formed on the edge.

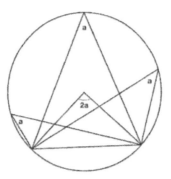

Example 20.5

In a regular octagon $ABCDEFGH$, find the measure of $\angle ACF$.

§20.3 Polygons

Essential Academy (June 2023) Introduction to AMC 8

> **Theorem 20.6**
>
> $$\text{Sum of Interior Angles of a Polygon} = 180(n-2)$$
> $$\text{Interior Angle of a Regular Polygon} = 180 \cdot \frac{n-2}{n}$$
> $$\text{Exterior Angle of a Regular Polygon} = \frac{360}{n}$$

Important Interior Angles

Number of Sides in Regular Polygon	Interior Angle of Regular Polygon
3	60
4	90
5	108
6	120
8	135
9	140
10	144

Example 20.7
(MATHCOUNTS) A square is located in the interior of a regular hexagon, and certain vertices are labeled as shown. What is the degree measure of $\angle ABC$?

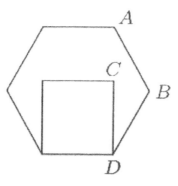

§20.4 Advanced Circle Angle Chasing Theorems

Theorem 20.8

The perpendicular bisector of any chord passes through the center. In the figure below, the perpendicular bisectors of AB and CD intersect at the center O.

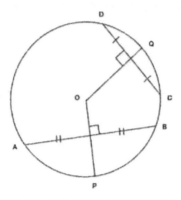

- Congruent chords are equidistant from the center of a circle.
- If two chords in a circle are congruent, then their intercepted arcs are congruent.
- If two chords in a circle are congruent, then they determine two central angles that are congruent.

Theorem 20.9

The angle marked in the diagram is half of the difference of the 2 red arcs.

$$\angle APC = \frac{\widehat{BD} - \widehat{AC}}{2}$$

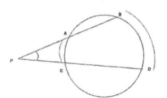

Theorem 20.10

If two chords AB and CD intersect at P, then the $\angle BPC$ and $\angle APD$ are equal to the average of the two arcs.

$$\angle BPC = \angle APD = \frac{\overset{\frown}{BC} + \overset{\frown}{AD}}{2}$$

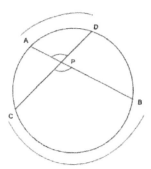

Theorem 20.11

If a tangent R intersects the circle at Q, and a chord QP is drawn, then the $\angle RQP$ is equal to half the arc angle.

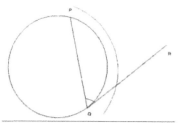

Theorem 20.12

(Equal Chords Mark Out Equal Arcs) If you have 2 chords of the same length, the sector of the circle they mark out will be equal.

Theorem 20.13
(Right Angle Tangency Point) If you connect the center of a circle to the point where the circle and a line are tangent, they will form a right angle.

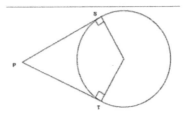

§20.5 Practice Problems

Problem 20.14
In the adjoining figure, $ABCD$ is a square, ABE is an equilateral triangle and point E is outside square $ABCD$. What is the measure of $\angle AED$ in degrees?.

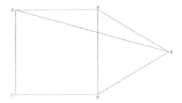

Problem 20.15
In quadrilateral $ABCD$, sides AB and BC both have length 10, sides CD and DA both have length 17, and the measure of angle ADC is $60°$. What is the length of diagonal AC?

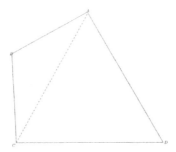

Problem 20.16

In $\triangle ABC$, D is a point on side AC such that $BD = DC$ and $\angle BCD$ measures $70°$. What is the degree measure of $\angle ADB$?

Problem 20.17

In the figure, point A is the center of the circle, the measure of angle RAS is $74°$, and the measure of angle RTB is $28°$. What is the measure of minor arc BR, in degrees?

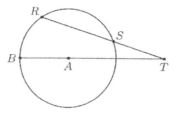

Problem 20.18

The circumference of the circle with center O is divided into 12 equal arcs, marked the letters A through L as seen below. What is the number of degrees in the sum of the angles x and y?

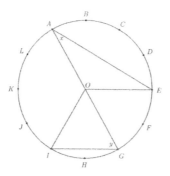

Problem 20.19

Two angles of an isosceles triangle measure $70°$ and $x°$. What is the sum of the different possible values of x?

Problem 20.20

Two congruent circles centered at points A and B each pass through the other circle's center. The line containing both A and B is extended to intersect the circles at points C and D. The circles intersect at two points, one of which is E. What is the degree measure of $\angle CED$?

Problem 20.21

The keystone arch is an ancient architectural feature. It is composed of congruent isosceles trapezoids fitted together along the non-parallel sides, as shown. The bottom sides of the two end trapezoids are horizontal. In an arch made with 9 trapezoids, let x be the angle measure in degrees of the larger interior angle of the trapezoid. What is x?

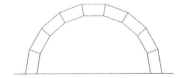

Problem 20.22

In triangle ABC, $AB = AC$ and $\angle A = 40°$. The bisector from $\angle B$ intersects AC at point D. What is $\angle BDC$?

Problem 20.23

Point P lies inside triangle ABC such that $\angle PBC = 30°$ and $\angle PAC = 20°$. If $\angle APB$ is a right angle, find the measure of $\angle BCA$ in degrees.

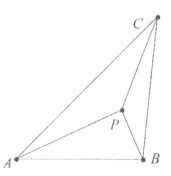

Problem 20.24

As shown in the figure below, point E lies on the opposite half-plane determined by line CD from point A so that $\angle CDE = 110°$. Point F lies on AD so that $DE = DF$, and $ABCD$ is a square. What is the degree measure of $\angle AFE$?

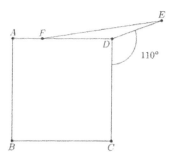

Problem 20.25

Concave quadrilateral $ABCD$ is symmetric about the line AC. The measures of angles DAB and ABC are 84 degrees and 32 degrees, respectively. The dashed line segments bisect angles ABC and ADC. What is the degree measure of the acute angle at which the two dashed line segments intersect?

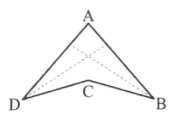

Problem 20.26

In triangle ABC, $\angle BAC$ is a right angle and $\angle ACB$ measures $34°$. Let D be a point on segment BC for which $AC = CD$, and let the angle bisector of $\angle CBA$ intersect line AD at E. What is the measure of $\angle BED$?

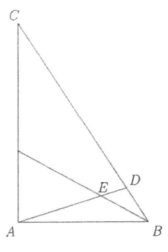

Problem 20.27

In square $ABCD$, point E lies on side BC and point F lies on side CD so that triangle AEF is equilateral and inside the square. Point M is the midpoint of segment EF, and P is the point other than E on AE for which $PM = FM$. The extension of segment PM meets segment CD at Q. What is the measure of $\angle CQP$, in degrees?

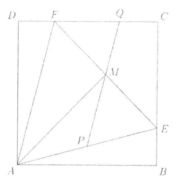

§21 Triangle

§21.1 Area of Triangle

Theorem 21.1
(Base and Height Formula) A triangle with base b and height h has an area of $\frac{1}{2}bh$.

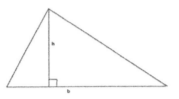

Theorem 21.2
(Heron Formula) The **semiperimeter** of a triangle with sides a, b, c. is defined to be $s = \frac{1}{2}(a+b+c)$. Then the area of the triangle equals
$$\sqrt{s(s-a)(s-b)(s-c)}$$

Example 21.3
(EMCC) Given that the heights of $\triangle ABC$ have lengths $\frac{15}{7}, 5$, and 3, what is the square of the area of $\triangle ABC$?

The inradius of a triangle is the radius of the inscribed circle in the triangle.

The incenter of a triangle is the intersection of all the angle bisectors. This point is also the center of the incircle, and equidistant from all the three sides.

Theorem 21.4
(Area of a Triangle Using Inradius) A triangle with inradius r (the radius of the circle that can be inscribed in a triangle) and semiperimeter s has an area of sr.

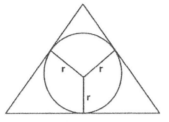

Theorem 21.5
(Inradius r of a Right Triangle)
$$r = \frac{1}{2}(a + b - c)$$
where a and b are the legs of the triangle, and c is the hypotenuse.

The **circumradius** of a triangle is the radius of circle that a triangle is inscribed in.

The **circumcenter** of a triangle is the intersection of all 3 perpendicular bisectors (the line that bisects a segment and is perpendicular to it). This point is also the center of the circumcircle and equidistant from all the three vertices.

Theorem 21.6

(Area of a Triangle Using Circumradius) A triangle with circumradius R (the radius of the circle that the triangle can be inscribed in) and sides a, b, c has an area of

$$\frac{abc}{4R}$$

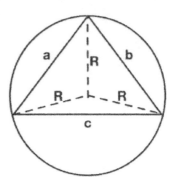

§21.2 Special Triangle

Theorem 21.7

(Equilateral Triangle) If the side length of an equilateral triangle is a

- Height of the Triangle $= \frac{\sqrt{3}}{2}a$
- Area of the Triangle $= \frac{\sqrt{3}}{4}a^2$

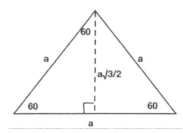

Theorem 21.8

$(45 - 45 - 90$ Triangle) If the side length of a $45 - 45 - 90$ triangle is a

- Hypotenuse of the Triangle $= \sqrt{2}a$
- Area of the Triangle $= \frac{1}{2}a^2$

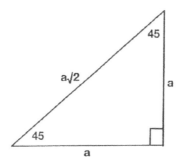

Theorem 21.9

$(30 - 60 - 90$ Triangle) If the short leg length of a $30 - 60 - 90$ triangle is a

- Long Leg of the Triangle $= \sqrt{3}a$
- Hypotenuse of the Triangle $= 2a$
- Area $= \frac{\sqrt{3}}{2}a^2$

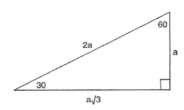

§21.3 Pythagorean Theorem

Theorem 21.10

A right triangle with legs a and b and hypotenuse c satisfies the following relation:
$$a^2 + b^2 = c^2$$

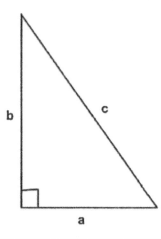

Theorem 21.11
(Special Properties of Right Triangles) In a right triangle ABC where B is the right angle, the following triangles are similar

$$\triangle ABC \sim \triangle ADB \sim \triangle BDC$$

which give the formulas

1. $AD \cdot CD = BD^2$
2. $AD \cdot AC = AB^2$
3. $CD \cdot CA = CB^2$

In addition, the length of the perpendicular to the hypotenuse $\overline{BD}$ is

$$\sqrt{\frac{AB \cdot BC}{AC}}$$

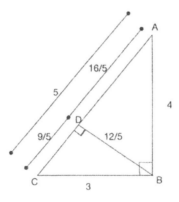

§21.4 Triangle Properties

A cevian is any line from any vertex of a triangle to the opposite side. Medians and angle bisectors are special cases of cevians.

A median is a line connecting a point to the midpoint of the opposite side.

In a triangle, the intersection of all 3 medians in a triangle is the centroid.

Theorem 21.12
The **centroid** of a triangle is on the median and it is $\frac{2}{3}$ of the way from from one of vertices to the midpoint of the opposite side.

Theorem 21.13

(Median in a Right Triangle) In a right triangle ABC, let the median from point B intersect AC at a point P. Then $AP = BP = CP$. Basically, in a right triangle AC is the diameter of the circumcircle, and PC, PA, and PB are radii of the circumcircle.

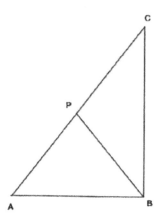

§21.5 Angle Bisector Theorem

Theorem 21.14

(Angle Bisector Theorem) If the line AD bisects angle A, then

$$\frac{AB}{BD} = \frac{AC}{CD}$$

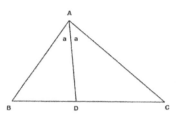

§21.6 Practice Problems

Problem 21.15
The twelve-sided figure shown has been drawn on 1 cm × 1 cm graph paper. What is the area of the figure in?

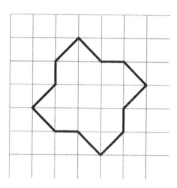

Problem 21.16
(AMC 8) A triangle with vertices as $A = (1,3)$, $B = (5,1)$, and $C = (4,4)$ is plotted on a 6×5 grid. What fraction of the grid is covered by the triangle?

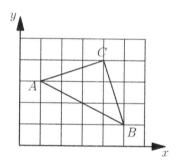

Problem 21.17
(AMC 8) Each of the following four large congruent squares is subdivided into combinations of congruent triangles or rectangles and is partially bolded. What percent of the total area is partially bolded?

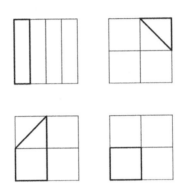

Problem 21.18
(AMC 8) In the figure below, choose point D on BC so that $\triangle ACD$ and $\triangle ABD$ have equal perimeters. What is the area of $\triangle ABD$?

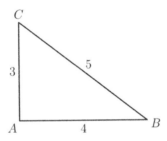

Problem 21.19
(AMC 8) In the non-convex quadrilateral $ABCD$ shown below, $\angle BCD$ is a right angle, $AB = 12$, $BC = 4$, $CD = 3$, and $AD = 13$. What is the area of quadrilateral $ABCD$?

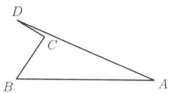

Problem 21.20
(AMC 8) The triangular plot of ACD lies between Aspen Road, Brown Road and a railroad. Main Street runs east and west, and the railroad runs north and south. The numbers in the diagram indicate distances in miles. The width of the railroad track can be ignored. How many square miles are in the plot of land ACD?

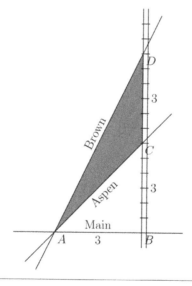

117

Problem 21.21

(AMC 8) In square $ABCE$, $\overline{AF} = 2\overline{FE}$ and $\overline{CD} = 2\overline{DE}$. What is the ratio of the area of $\triangle BFD$ to the area of square $ABCE$?

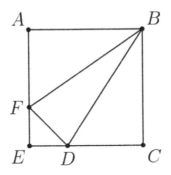

Problem 21.22

(AMC 8) Rectangle $ABCD$ and right triangle DCE have the same area. They are joined to form a trapezoid, as shown. What is DE?

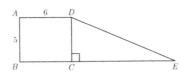

Problem 21.23

What is the area of the shaded pinwheel shown in the 5×5 grid?

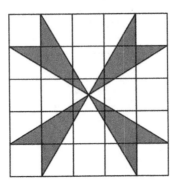

Problem 21.24
What is the area of the triangle formed by the lines $y = 5, y = 1 + x$, and $y = 1 - x$?

Problem 21.25
In triangle ABC, point D divides side AC so that $\overline{AD} : \overline{DC} = 1 : 2$. Let E be the midpoint of BD and let F be the point of intersection of line BC and line AE. Given that the area of $\triangle ABC$ is 360, what is the area of $\triangle EBF$?

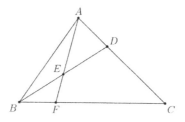

§22 Quadrilateral

§22.1 Square

Any square with side length s has area s^2 and perimeter $4s$.

> **Example 22.1**
> A large square region is paved with n^2 gray square tiles, each measuring s inches on a side. A border d inches wide surrounds each tile. The figure below shows the case for $n = 3$. When $n = 24$, the 576 gray tiles cover 64% of the area of the large square region. What is the ratio $\frac{d}{s}$ for this larger value of n?
>
>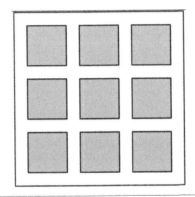

§22.2 Rectangle

Area of Rectangle:

§22.3 Rhombus

Area of Rhombus:

§22.4 Parallelogram

Area of Parallelogram:

§22.5 Trapezoid

Area of Trapezoid:

> **Example 22.2**
> (EMCC) In quadrilateral $PEAR$, $PE = 21$, $EA = 20$, $AR = 15$, $RE = 25$, and $AP = 29$. Find the area of the quadrilateral.

§22.6 Practice Problems

> **Problem 22.3**
> (AMC 8) Three identical rectangles are put together to form rectangle $ABCD$, as shown in the figure below. Given that the length of the shorter side of each of the smaller rectangles is 5 feet, what is the area in square feet of rectangle $ABCD$?
>
>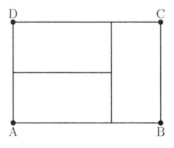

Problem 22.4

(AMC 8) Points A, B, C and D are midpoints of the sides of the larger square. If the larger square has area 60, what is the area of the smaller square?

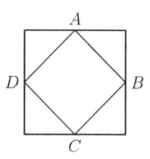

Problem 22.5

(AMC 8) Quadrilateral $ABCD$ is a trapezoid, $AD = 15, AB = 50, BC = 20$, and the altitude is 12. What is the area of the trapezoid?

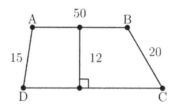

Problem 22.6

(AMC 8) The midpoints of the four sides of a rectangle are $(-3, 0), (2, 0), (5, 4)$, and $(0, 4)$. What is the area of the rectangle?

§23 Circles

§23.1 Circle Theorem

Area and Circumference

A circle with radius r has

Arcs of a Circle

§23.2 Circular Area

Example 23.1
(EMCC) Four congruent semicircles are drawn within the boundary of a square with side length 1. The center of each semicircle is the midpoint of a side of the square. Each semicircle is tangent to two other semicircles. Region R consists of points lying inside the square but outside of the semicircles. The area of R can be written in the form $a - b\pi$, where a and b are positive rational numbers. Compute $a + b$.

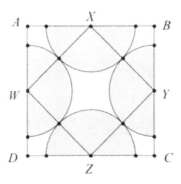

Example 23.2

(EMCC) Suppose points A and B lie on a circle of radius 4 with center O, such that $\angle AOB = 90°$. The perpendicular bisectors of segments OA and OB divide the interior of the circle into four regions. Find the area of the smallest region.

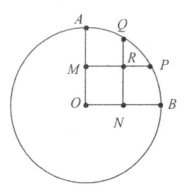

§23.3 Length Inside Circles

Example 23.3

(AMC 8) Rectangle $ABCD$ is inscribed in a semicircle with diameter FE, as shown in the figure. Let $DA = 16$, and let $FD = AE = 9$. What is the area of $ABCD$?

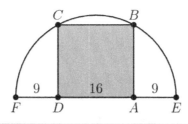

Example 23.4

A circle, centered at C, of radius of 25 has 2 parallel chords with lengths shown. To the nearest integer, what is the distance between these chords?

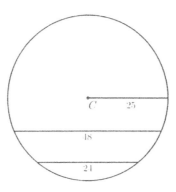

§23.4 Practice Problems

Problem 23.5

(AMC 8) In the diagram below, a diameter of each of the two smaller circles is a radius of the larger circle. If the two smaller circles have a combined area of 1 square unit, then what is the area of the shaded region, in square units?

Problem 23.6

(AMC 8) The two circles pictured have the same center C. Chord AD is tangent to the inner circle at B, AC is 10, and chord AD has length 16. What is the area between the two circles?

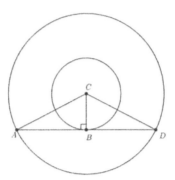

Problem 23.7

(AMC 8) A circle with radius 1 is inscribed in a square and circumscribed about another square as shown. Which fraction is closest to the ratio of the circle's shaded area to the area between the two squares?

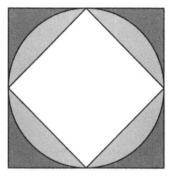

Problem 23.8

(AMC 8) Rectangle $ABCD$ has sides $CD = 3$ and $DA = 5$. A circle with a radius of 1 is centered at A, a circle with a radius of 2 is centered at B, and a circle with a radius of 3 is centered at C. Which of the following is closest to the area of the region inside the rectangle but outside all three circles?

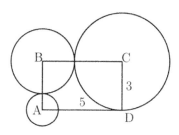

Problem 23.9

(AMC 8) Isosceles right triangle ABC encloses a semicircle of area 2π. The circle has its center O on hypotenuse AB and is tangent to sides AC and BC. What is the area of triangle ABC?

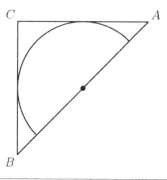

Problem 23.10

(AMC 8) Margie's winning art design is shown. The smallest circle has radius 2 inches, with each successive circle's radius increasing by 2 inches. Which of the following is closest to the percent of the design that is black?

Problem 23.11

(AMC 8) Semicircles POQ and ROS pass through the center of circle O. What is the ratio of the combined areas of the two semicircles to the area of circle O?

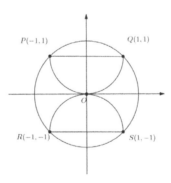

Problem 23.12

(AMC 8) A circle of radius 2 is cut into four congruent arcs. The four arcs are joined to form the star figure shown. What is the ratio of the area of the star figure to the area of the original circle?

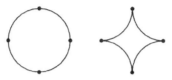

Problem 23.13

(AMC 8) In the right triangle ABC, $AC = 12$, $BC = 5$, and angle C is a right angle. A semicircle is inscribed in the triangle as shown. What is the radius of the semicircle?

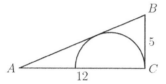

Problem 23.14

(AMC 8) A semicircle is inscribed in an isosceles triangle with base 16 and height 15 so that the diameter of the semicircle is contained in the base of the triangle as shown. What is the radius of the semicircle?

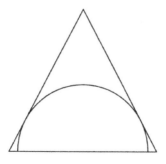

Problem 23.15

(AMC 8) In the figure shown, US and UT are line segments each of length 2, and $\angle TUS = 60°$. Arcs TR and SR are each one-sixth of a circle with radius 2. What is the area of the region shown?

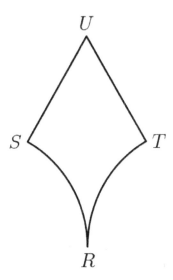

Problem 23.16

(AMC 10) Three equally spaced parallel lines intersect a circle, creating three chords of lengths $38, 38$, and 34. What is the distance between two adjacent parallel lines?

Problem 23.17

A rectangle of height 10 and width 24 is inscribed in a circle. What is the circumference of that circle? Express your answer in terms of π.

§24 Similarity and Congruence

§24.1 Congruent Triangle

Two triangles are congruent if the three angles and sides in the triangle are the same. In other words, the triangles are the same (not necessarily same orientation).

In general, triangles are congruent if:

- AAS congruence: Two angles of the triangles are same and the 1 side next to the 2 angles are equal
- SAS Congruence (Side Angle Side): Two sides are equal and the angle between the sides are equal
- SSS congruence (Side Side Side): All three sides are congruent
- HL congruence (Hypotenuse Leg): In a right triangle, hypotenuse and leg are equal
- LL congruence (LL Leg): In a right triangle, the two legs are equal

Example 24.1
In square $ABCD$, points E and F lie on segments AD and CD, respectively. Given that $\angle EBF = 45°$, $DE = 12$, and $DF = 35$, compute AB.

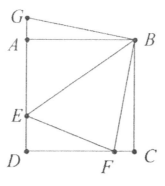

§24.2 Similar Triangle

In general, triangles are similar if:

- AA similarity: Two angles of the triangles are same, which basically means that the third angle will be equal)
- SAS similarity (Side Angle Side): Two sides are proportional and the angle between the sides is

equal

- SSS similarity (Side Side Side): All three sides are proportional
- HL similarity (Hypotenuse Leg): In a right triangle, the hypotenuse and leg are proportional
- LL similarity (LL Leg): In a right triangle, the two legs are proportional

Lemma 24.2

For similar triangles:

1. All the angles of the triangles are same
2. All corresponding sides have same ratio
3. Area ratio is the square of side length ratio

Example 24.3

(EMCC) In triangle $ABC, \angle BAC = 90°$, and the length of segment AB is 2011. Let M be the midpoint of BC and D the midpoint of AM. Let E be the point on segment AB such that $EM \parallel CD$. What is the length of segment BE?

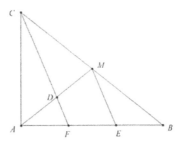

Example 24.4

(EMCC) In triangle ABC with $BC = 5, CA = 13$, and $AB = 12$, Points E and F are chosen on sides AC and AB, respectively, such that $EF \parallel BC$. Given that triangle AEF and trapezoid $EFBC$ have the same perimeter, find the length of EF.

§24.3 Practice Problems

Problem 24.5

(AMC 8) The diagram shows an octagon consisting of 10 unit squares. The portion below $\overline{PQ}$ is a unit square and a triangle with base 5. If $\overline{PQ}$ bisects the area of the octagon, what is the ratio $\dfrac{XQ}{QY}$?

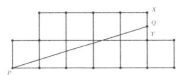

Problem 24.6

(AMC 8) Rectangle $DEFA$ below is a 3×4 rectangle with $DC = CB = BA = 1$. The area of the "bat wings" (shaded area) is

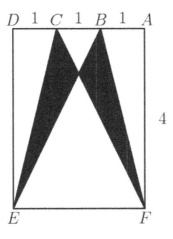

Problem 24.7

(AMC 8) One-inch squares are cut from the corners of this 5 inch square. What is the area in square inches of the largest square that can be fitted into the remaining space?

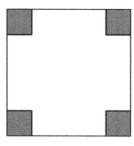

§25 Area of Polygon

§25.1 Area of Polygon

Hexagon

> **Theorem 25.1**
> Properties of Hexagon:
>
> - Sum of Interior Angle of a Regular Hexagon $= (6-2) \cdot 180 = 720$
> - Interior Angle of a Regular Hexagon $= \frac{(6-2)}{6} \cdot 180 = 120$
> - Exterior Angle of a Regular Hexagon $= \frac{360}{6} = 60$
> - Area of a Regular Hexagon $= 6 \cdot \frac{\sqrt{3}}{4} a^2$
> - Length of the Diagonal of a Regular Hexagon $= 2a$
>
>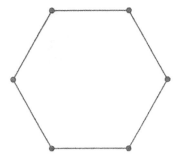

Octagon

Theorem 25.2

Properties of Octagon:

- Sum of Interior Angle of a Regular Octagon $= (8-2) \cdot 180 = 1080$
- Interior Angle of a Regular Octagon $= \frac{(8-2)}{8} \cdot 180 = 135$
- Exterior Angle of a Regular Octagon $= \frac{360}{8} = 45$
- Area of a Regular Octagon $= 2(1+\sqrt{2})^2 s^2$

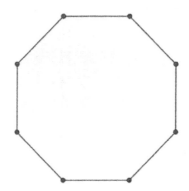

§25.2 Practice Problems

Problem 25.3

(AMC 8) Point O is the center of the regular octagon $ABCDEFGH$, and X is the midpoint of the side AB. What fraction of the area of the octagon is shaded?

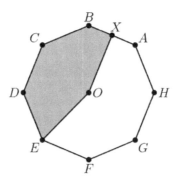

Problem 25.4

(AMC 8) In the figure, the outer equilateral triangle has area 16, the inner equilateral triangle has area 1, and the three trapezoids are congruent. What is the area of one of the trapezoids?

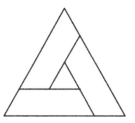

Problem 25.5

(AMC 8) The letter T is formed by placing two 2×4 inch rectangles next to each other, as shown. What is the perimeter of the T, in inches?

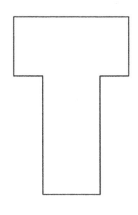

Problem 25.6

(AMC 8) A unit hexagram is composed of a regular hexagon of side length 1 and its 6 equilateral triangular extensions, as shown in the diagram. What is the ratio of the area of the extensions to the area of the original hexagon?

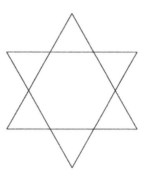

Problem 25.7

(AMC 8) The area of polygon $ABCDEF$ is 52 with $AB = 8, BC = 9$ and $FA = 5$. What is $DE + EF$?

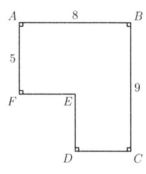

Problem 25.8

(AMC 8) A decorative window is made up of a rectangle with semicircles on either end. The ratio of AD to AB is $3:2$, and AB is 30 inches. What is the ratio of the area of the rectangle to the combined areas of the semicircles?

Problem 25.9

(AMC 8) Triangle ABC is an isosceles triangle with $AB = BC$. Point D is the midpoint of both BC and AE, and CE is 11 units long. Triangle ABD is congruent to triangle ECD. What is the length of BD?

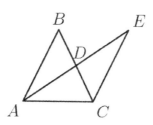

Problem 25.10

(AMC 8) What is the perimeter of trapezoid $ABCD$?

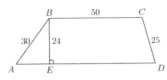

§26 3D Geometry

§26.1 Cube

Theorem 26.1
Properties of Cube with Side Length a:

- Volume of a Cube $= a^3$
- Surface Area of Cube $= 6a^2$
- Length of Space Diagonal of a Cube $= \sqrt{3}a$

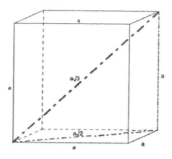

§26.2 Prism

Theorem 26.2
Properties of Prism with lengths l, b, h :

- Volume of Rectangular Prism $= lbh$
- Surface Area of Rectangular Prism $= 2(lb + bh + lh)$
- Length of Space Diagonal of Rectangular Prism $= \sqrt{l^2 + b^2 + h^2}$

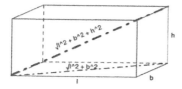

Example 26.3

(AMC 8) The figure below shows a polygon $ABCDEFGH$, consisting of rectangles and right triangles. When cut out and folded on the dotted lines, the polygon forms a triangular prism. Suppose that $AH = EF = 8$ and $GH = 14$. What is the volume of the prism?

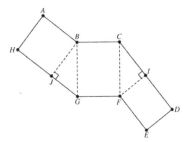

§26.3 Pyramid

Theorem 26.4

$$\text{Volume of Pyramid} = \frac{1}{3} \cdot \text{base area} \cdot \text{height}$$

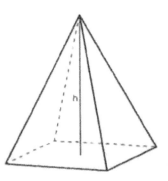

Theorem 26.5

(Volume of Regular Square Pyramid (all sides equal))

$$\text{Volume of Regular Pyramid} = \frac{\sqrt{2}}{6}s^3$$

Theorem 26.6
(Volume of Regular Tetrahedron (all sides equal))

$$\text{Volume of Regular Tetrahedron} = \frac{\sqrt{2}}{12}s^3$$

§26.4 Cylinder

Theorem 26.7
(Volume and Surface Area of a Cylinder)

$$\text{Volume of a Cylinder} = \pi r^2 h$$

$$\text{Surface Area of a Cylinder} = 2\pi r^2 + 2\pi rh$$

§26.5 Cone

Theorem 26.8
(Volume and Surface Area of a Cone)

$$\text{Volume of a Cone} = \frac{1}{3}\pi r^2 h$$

$$\text{Surface Area of a Cone} = \pi r^2 + \pi rl$$

where L is the lateral or slant height.

§26.6 Sphere

Theorem 26.9
(Volume and Surface Area of a Sphere)

$$\text{Volume of a Sphere} = \frac{4}{3}\pi r^3$$

$$\text{Surface Area of a Sphere} = 4\pi r^2$$

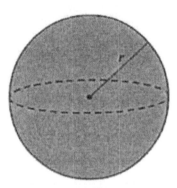

§26.7 Practice Problems

Problem 26.10
(AMC 8) Alex and Felicia each have cats as pets. Alex buys cat food in cylindrical cans that are 6 cm in diameter and 12 cm high. Felicia buys cat food in cylindrical cans that are 12 cm in diameter and 6 cm high. What is the ratio of the volume of one of Alex's cans to the volume one of Felicia's cans?

Problem 26.11

(AMC 8) Isabella uses one-foot cubical blocks to build a rectangular fort that is 12 feet long, 10 feet wide, and 5 feet high. The floor and the four walls are all one foot thick. How many blocks does the fort contain?

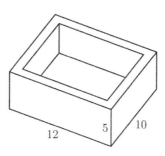

Problem 26.12

(AMC 8) Jerry cuts a wedge from a 6-cm cylinder of bologna as shown by the dashed curve. Which answer choice is closest to the volume of his wedge in cubic centimeters?

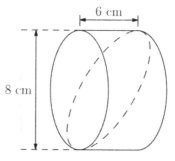

Problem 26.13

(AMC 8) A square with area 4 is inscribed in a square with area 5, with one vertex of the smaller square on each side of the larger square. A vertex of the smaller square divides a side of the larger square into two segments, one of length a, and the other of length b. What is the value of ab?

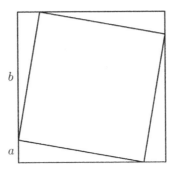

Problem 26.14

(AMC 8) In the cube $ABCDEFGH$ with opposite vertices C and E, J and I are the midpoints of segments $\overline{FB}$ and $\overline{HD}$, respectively. Let R be the ratio of the area of the cross-section $EJCI$ to the area of one of the faces of the cube. What is R^2?

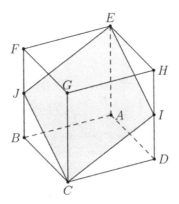

Problem 26.15

The tiny island nation of Turkey is a cone with height 12 meters and base radius 9 meters, with the base of the cone at sea level. If the sea level rises 4 meters, what is the surface area of Turkey that is still above water, in square meters?

Problem 26.16

A white cylindrical silo has a diameter of 30 feet and a height of 80 feet. A red stripe with a horizontal width of 3 feet is painted on the silo, as shown, making two complete revolutions around it. What is the area of the stripe in square feet?

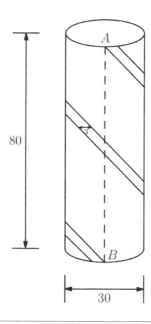

Introduction to AMC 8: Solutions Manual

ESSENTIAL ACADEMY

June 2023

Contents

I. Combinatorics 4

1 Counting 5

2 Permutations 6

3 Combinations 8

4 Probability 10

5 Casework Problems 13

6 Principles of Inclusion and Exclusion 14

7 Stars and Bars 17

8 Geometric Counting 19

9 Recursion 21

II. Algebra 24

10 Ratio and Percentage 25

11 Linear Function and Quadratic Function 28

12 Speed, Distance, and Time 33

13 Sequence and Series 36

14 Mean, Median, and Mode 38

15 Telescoping 42

III. Number Theory 44

16 Primes and Divisibility 45

17 Factors 48

18 GCF and LCM 50

19 Modular Arithmetic 51

IV. Geometry 52

20 Angle Chasing 53

21 Triangle 55

22 Quadrilateral 57

23 Circles	**58**
24 Similarity and Congruence	**61**
25 Area of Polygon	**62**
26 3D Geometry	**63**

Combinatorics

§1 Counting

1.1 $\boxed{61}$

1.2 $\boxed{60}$

1.3 $\boxed{33}$

1.4 $\boxed{49}$

1.5
$$200 - 198 + 196 - 194 + \cdots + 4 - 2 = (200 - 198) + (196 - 194) + \cdots + (4 - 2) = 2 \cdot 50 = \boxed{100}$$

1.6
$$-1 + 2 - 3 + 4 - 5 + 6 - 7 + \cdots - 499 + 500 = (-1 + 2) + (-3 + 4) + \cdots + (-499 + 500) = 1 \cdot 250 = \boxed{250}$$

Essential Academy (June 2023) Introduction to AMC 8: Solutions Manual

§2 Permutations

2.1 $\boxed{1000}$

2.4 $\boxed{24}$

2.5 There are 4 spots for the first book, 3 spots for the second book, 2 spots for the third book, and 1 spot for the fourth book. Hence, there are a total of $4! = \boxed{24}$. arrangements.

For the AMC8 videos, the same applies. There are 4! possible ways to order the 4 AMC videos, so the maximum number of students is $\boxed{24}$.

2.6 There are 3 spots in the line. There are 5 choices for the first spot, 4 choices for the second, and 3 choices for the third. Therefore, our answer is $5 \times 4 \times 3 = \boxed{60}$.

2.8 1. $\boxed{210}$, 2. $\boxed{20}$

2.9 $\boxed{210}$

2.10 $\boxed{4536}$

2.11 $\boxed{300}$

2.12 The first and fifth, second and fourth numbers of our number have the be the same. There are 9 was to pick the first number ($1 \sim 9$ excluding 0) and 10 ways to pick the second. The third number is free so there are 10 ways to pick it. The fourth and fifth numbers would be automatically assigned. Therefore, our answer is $9 \times 10 \times 10 = \boxed{900}$.

2.13 Clearly, the third digit has to 7. This leaves us to choose the first two numbers from the set $\{6, 8, 9\}$ which gives $3 \times 2 = 6$ ways. The last two numbers need to be chosen from $0 \sim 5$ which gives $6 \times 5 = 30$ ways. The fourth digit (which has no constraints) will have $10 - 1 - 2 - 2 = 5$ choices to choose from. Therefore, our answer is $6 \times 30 \times 5 = \boxed{900}$.

2.14 $\boxed{24}$

2.15 Interpret the problem as if Alex and Bob are tied together. Then, there are 4 separate entities we need to arrange around a circle which gives $\frac{4!}{4} = 6$ ways. However, Alex and Bob can switch spots (AB or BA clockwise) which gives in total $6 \times 2 = \boxed{12}$ ways.

2.16 $6 \cdot 5 \cdot 4 = \boxed{120}$

2.17 $5 \cdot 4 \cdot 3 \cdot 2 \cdot 1 = \boxed{120}$

2.18 $8 \cdot 7 \cdot 6 = \boxed{336}$

2.19 $_7P_5 = 7 \cdot 6 \cdot 5 \cdot 4 \cdot 3 = \boxed{2520}$

2.20 $\frac{8!}{2!2!} = \boxed{10080}$

2.21 $\frac{6!}{2!} = \boxed{360}$

2.22 $\frac{6!}{2!} = \boxed{360}$

2.23 $\frac{6!}{3!2!1!} = \boxed{60}$

2.24 $6! \cdot 2! = \boxed{1440}$

2.25 $5! = \boxed{120}$

§3 Combinations

3.1 $\boxed{20}$

3.3 $_{10}C_7 = \boxed{120}$.

3.4 After choosing the 2 required chapters, all that is left is to choose 5 chapters from the remaining 8 non-required chapters. $_8C_5 = \boxed{56}$.

3.6 $\boxed{32}$.

3.8 For each pencil or eraser we either have to choices: take it or leave it. Therefore, our answer is $2^8 = \boxed{256}$.

3.9 $\boxed{4950}$.

3.10 There are $_{12}C_3$ ways to choose the Alice's bars, which leaves 9 bars. Then we'll have $_9C_4$ ways to choose Betty's bars. The remaining bars will go to Chase. Therefore, our answer is
$$_{12}C_3 \cdot {_9}C_4 = \boxed{27720}$$

3.11 There are 4×6 choices for the first ball. After the first ball has been filled, there are 3×5 choices for the second ball, as the available row and column shrinks by 1. For the third ball, there are 2×4 choices. Finally, for the last ball, there are 1×3 choices. Therefore, our answer is
$$(4 \times 6)(3 \times 5)(2 \times 4)(1 \times 3) = \boxed{8640}.$$

3.12
$$\frac{9!}{2!2!} = \boxed{90720}$$

3.13
$$\frac{8!}{3!2!2!} = \boxed{1680}$$

3.14
$$\frac{5!}{2!} = \boxed{60}$$

3.16 $\boxed{2880}$.

3.17 Group L, O, I together and view them as one. We have two configurations (LOI) or (IOL). Other than this block, we have $2P, 2L, 1O$. Therefore, the number of ways to arrange these letters are
$$2 \cdot \frac{6!}{2!2!1!1!} = \boxed{360}.$$

3.18 A pentagon requires 5 lines, each in 5 different directions. For each distinct direction, there are 3 choices for lines. Therefore, the number of choices we have in total is $3^5 = \boxed{243}$.

3.19 Since the path must be ten-units, we can only move right or up. We have to move 6 right and 4 up in total.

To use complementary counting, let's count the number of ten-unit paths between A and B that cross X. The number of ways to go from A to X is $_4C_2 = 6$ as that is the number of ways to arrange

$2R, 2U$. Similarly, the number of ways to go from X to B is $_6C_2 = 15$. Hence, there are $6 \cdot 15 = 90$ ways to go from A to X to B.

Now let's subtract this from the total number of ways to go from A to B. The total number of ways to go from A to B is $_{10}C_4 = 210$. Therefore, our final answer is $210 - 90 = \boxed{120}$.

3.20 Let's count the number of vertices. For every pair of lines there is one intersection, so the number of vertices is $_5C_2 = 10$. Now using these 10 vertices, we can choose 3 of them to form a triangle. Therefore, the number of triangles that can be formed is $_{10}C_3 = \boxed{120}$.

3.21 The number of ways John can walk from the starting point to the deli is the number of distinct orderings of $3W, 3N$, which is equal to $_6C_3 = 20$. He also has the same 20 paths to choose from on his way returning. Therefore, our answer is $20^2 = \boxed{400}$.

§4 Probability

4.2 There are 3 primes from $1 \sim 6$. So, our answer is $\frac{3}{6} = \boxed{\frac{1}{2}}$.

4.3 Prime factorizing 648, we get $2^6 \cdot 23$. The product of our 4 dice cannot be a multiple of 23 as no number from $1 \sim 6$ is divisible by 23. Therefore, the product of our 4 dice cannot be a multiple of 648, so our probability is $\boxed{0}$.

4.4 $\boxed{\dfrac{13}{50}}$.

4.5 At least one of the two dices that are rolled have to be a multiple of 3. Let's split cases based on whether person B choose 18 or not as 18 is the only multiple of 3 in the second dice.

Case 1: B chooses 18.
If B chooses $18 = 2 \cdot 3^2$ we need a multiple of 4 from person A. However, person A has no multiple of 4, so there are 0 desired outcomes here.

Case 2: B does not choose 18.
If B does not choose 18, we need a multiple of 3 from person A. If Person A chooses 6 then we need a multiple of 4 from person B: 4, 28, 44 work. If Person A chooses 15 then we need a multiple of 8 from person B: none work. If Person A chooses 30 then we need a multiple of 4 from person B: 4, 28, 44 work. Therefore, the total number of desired outcomes is 6, giving a probability of $\boxed{\dfrac{1}{6}}$.

4.6 There are a total of $7! = 5040$ possible outcomes. To count the number of desirable outcomes, fix Charlie in first place and George in sixth place. Then, we have $5!$ ways to order the remaining people. Therefore, our final answer is $\frac{5!}{7!} = \boxed{\dfrac{1}{42}}$.

4.7 The probability it will rain on both days is

$$P(\text{rain on saturday}) \cdot P(\text{rain on sunday} \mid \text{rain on saturday}) = 0.2 \cdot 0.3 = 0.06$$

and the probability it will not rain on both days is

$$P(\text{no rain on saturday}) \cdot P(\text{no rain on sunday} \mid \text{no rain on saturday}) = (1 - 0.2) \cdot (1 - 0.1) = 0.72$$

In total, our probability is $0.06 + 0.72 = \boxed{0.78}$.

4.8 Let x be the probability that it rains on Sunday given that it doesn't rain on Saturday. We then have

$$\frac{3}{5}x + \frac{2}{5}2x = \frac{3}{10} \implies \frac{7}{5}x = \frac{3}{10} \implies x = \frac{3}{14}$$

. Therefore, the probability that it doesn't rain on either day is

$$\left(1 - \frac{3}{14}\right)\left(\frac{3}{5}\right) = \frac{33}{70}$$

. Therefore, the probability that rains on at least one of the days is $1 - \frac{33}{70} = \frac{37}{70}$, so adding up the 2 numbers, we have $37 + 70 = \boxed{107}$.

4.10 The probability there will be no rain at all in the next 5 days is 0.75^5. Hence, the probability it will rain at least once in the next five days is $1 - 0.75^5 = \boxed{\dfrac{781}{1024}}$.

4.11 Out of a total $6 \times 6 = 36$ possible outcomes, the number of ways for the sum to be greater than 11 is 1, namely $(6,6)$. Therefore, the number of ways for the sum to be 11 or less is $\boxed{35}$.

4.12 Choose the odd numbers from the subset $\{1,3,5,7\}$ and the even numbers from the subset $\{2,4,6\}$. The number of ways to choose 1 or 0 odd numbers is $1 + 4 = 5$, so the number of ways to choose at least 2 odd numbers is $2^4 - 5 = 11$. The number of ways to choose 0 even numbers is 1, so the number of ways to choose at least 1 even number is $2^3 - 1 = 7$. Therefore, in total, we have $\boxed{77}$ possible ways.

4.13 The probability of an even number appearing is $\frac{3}{4}$, while the probability of an odd number appearing is $\frac{1}{4}$. Then the probability of getting an odd and an even (to make an odd number) is $\frac{3}{4} \cdot \frac{1}{4} \cdot 2 = \frac{3}{8}$. Then the probability of getting an even number is $1 - \frac{3}{8} = \boxed{(E) \ \dfrac{5}{8}}$.

4.14 As $1, 2, 3$ are the number of outcomes that give sums $2, 3, 4$, the probability that the sum of the numbers is 4 or less is $\frac{1+2+3}{36} = \frac{1}{6}$. Hence, the probability the sum of the numbers is 5 or more is $1 - \frac{1}{6} = \boxed{\dfrac{5}{6}}$.

4.15 The number of subsets that have at least 1 even number. From a total of $2^{10} = 1024$ we exclude the number of subsets that have no even numbers. The number of subsets that have no even numbers is $2^5 = 32$. Hence, the number of subsets that have at least 1 even number is $1024 - 32 = 992$. Subtracting the case when all the elements are contained our desired answer is $992 - 1 = \boxed{991}$.

4.16 For the first digit, we have 9 choices as can choose from $1 \sim 9$. The second digit also has 9 choices as it can be anything from $0 \sim 9$ except the one chosen for the first digit. Therefore, our answer is $9 \cdot 9 = \boxed{81}$.

4.17 There are in total $_7C_4 = 35$ ways to assign parking spaces without constraint. There is 1 way of assigning parking spaces with no two cars being adjacent is 1 (positions $1, 3, 5, 7$). Using complementary counting, our answer is $35 - 1 = \boxed{34}$.

4.18 There are in total 6! ways to line up the 6 people. Let's subtract the cases when Messi and Ronaldo stand next to each other. Grouping Messi and Ronaldo into one, we will have 5! ways to arrange the group (MR) with 4 others. Messi and Ronaldo can switch spots, which gives 2 ways. Therefore, our answer is $2 \cdot 5! = \boxed{240}$.

4.19 Let n be the number of positive integers less than 50 that are divisible by 3 or 5. There are 16 numbers that are divisible by 3 in this range, 9 numbers that are divisible by 5, and 3 numbers that are divisible by 15. Therefore, $n = 16 + 9 - 3 = 22$. Subtracting this number from our original total, we get $49 - 22 = \boxed{27}$.

4.20 Let n be the number of orderings to flip more heads than tails. By symmetry, the number of orderings to flip more heads than tails is the same as the number of orderings to flip more tails than heads. The number of orderings to flip an equal number of heads and tails is the number of ways to order $5H, 5T$, which is $_{10}C_5 = 252$. Now we can set up the equation

$$2n + 252 = 2^{10} = 1024$$

as the three cases are collectively exhaustive. $\therefore n = \boxed{386}$.

4.21 Let's count the number of vertices. For every pair of lines there is one intersection, so the number of vertices is $_5C_2 = 10$. Now using these 10 vertices, we can choose 3 of them to form a triangle. Therefore, the number of triangles that can be formed is $_{10}C_3 = \boxed{120}$.

§5 Casework Problems

5.1 To reach a sum of 3 we either roll 2, 1 or 1, 2. $\boxed{2}$.

5.2 We can split into three cases:

1. Both cards are green: $_4C_2 = 6$.

2. Both cards are red: $_4C_2 = 6$.

3. Both cards have the same number: $_4C_1 = 4$.

These three cases are mutually exclusive, so there is no overcounting. In total, we have $\boxed{16}$.

5.3 Without loss of generality, assume we picked up a green ball from the first box. Then, in order to find another green ball from another box we choose a different box with probability $\frac{9}{10}$ and a green ball from that different box with probability $\frac{1}{4}$. Therefore, our answer is

$$\frac{9}{10} \cdot \frac{1}{4} = \boxed{\frac{9}{40}}$$

5.4 $\boxed{26}$.

Essential Academy (June 2023) — Introduction to AMC 8: Solutions Manual

§6 Principles of Inclusion and Exclusion

6.1 $\boxed{57}$.

6.2 $\boxed{872}$.

6.3 $\boxed{138}$.

6.4 There are total of 3 multiples of 7 from $1 \sim 25$ and 22 numbers not divisible by 7. The total number of outcomes Mark can guess from is $3 \cdot 22 = 66$. Without loss of generality, let's say George chose 1 as the non-multiple and 7 as the number divisible by 7.

Now the number of outcomes where Mark guesses the multiple of 7 correctly is 22 as he has 22 choices for the non-multiple. Similarly, the number of outcomes where Mark guesses the non-multiple of 7 is 3. However, we would have overcounted one case when Mark guesses both 1 and 7. By Principle of Inclusion and Exclusion, the total number of desirable outcomes is $22 + 3 - 1 = 24$. Hence, our answer is

$$\frac{24}{66} = \boxed{\frac{4}{11}}$$

6.5 We have 3 Ls and 2 Os, so we must have the configuration $LOLOL$. Thinking of $LOLOL$ as a chunk, and arranging it along with $2P$ and $1I$, we have a total of $\frac{4!}{2!1!1!} = \boxed{12}$ ways.

6.8 Without the restriction that each student receives at least one award, we could simply take each of the 5 awards and choose one of the 3 students to give it to, so that there would be $3^5 = 243$ ways to distribute the awards. We now need to subtract the cases where at least one student doesn't receive an award. If a student doesn't receive an award, there are 3 choices for which student that is, then $2^5 = 32$ ways of choosing a student to receive each of the awards, for a total of $3 \cdot 32 = 96$. However, if 2 students both don't receive an award, then such a case would be counted twice among our 96, so we need to add back in these cases. Of course, 2 students both not receiving an award is equivalent to only 1 student receiving all 5 awards, so there are simply 3 choices for which student that would be. Therefore, the total number of ways of distributing the awards is $243 - 96 + 3 = \boxed{150}$.

6.10 Using PIE, our answer is $50 + 33 - 16 = \boxed{67}$.

6.11 This problem complements the previous problem. Subtract the answer from **6.10** to get $100 - 67 = \boxed{33}$.

6.12 By PIE, the number of students taking Geometry or Algebra is $100 + 70 - 30 = 140$. Therefore, the number of students taking neither is $200 - 140 = \boxed{60}$.

6.13 Simply subtract the number of adults who own motorcycles from the total. $351 - 45 = \boxed{306}$.

6.14 Draw a Venn Diagram with two sets A, B. Set A represents the students who voted in favor of the first issue and set B represents the students who voted in favor of the second issue. $A \cap B$ would each represent students who voted in favor of both issues. Students who voted against both issues would be outside the two sets making up the Venn Diagram. . The problem gives us that $|A| = 149, |B| = 119$, and $|A \cup B| = 198 - 29 = 169$. Therefore, our answer is

$$|A \cap B| = |A| + |B| - |A \cup B| = 149 + 119 - 169 = \boxed{99}.$$

6.15 Let A, B, C be the set of multiples of $2, 3, 5$ among the positive numbers less than or equal to

200. Interpreting the problem using a Venn Diagram, we wish to find $|A \cap B| - |A \cap B \cap C|$. $|A \cap B|$ counts all the multiples of 6 from $1 \sim 200$, which is 33. $|A \cap B \cap C|$ counts all the multiples of 30 from $1 \sim 200$, which is 6. Therefore, our answer is $33 - 6 = \boxed{27}$.

6.16 By PIE (Property of Inclusion/Exclusion), we have

$$|A_1 \cup A_2 \cup A_3| = \sum |A_i| - \sum |A_i \cap A_j| + |A_1 \cap A_2 \cap A_3|.$$

Number of people in at least two sets is $\sum |A_i \cap A_j| - 2|A_1 \cap A_2 \cap A_3| = 9$. So, $20 = (10+13+9)-(9+2x)+x$, which gives $x = \boxed{3}$. **6.17** We use Principle of Inclusion-Exclusion. There are 365 days in the year, and we subtract the days that she gets at least 1 phone call, which is

$$\left\lfloor \frac{365}{3} \right\rfloor + \left\lfloor \frac{365}{4} \right\rfloor + \left\lfloor \frac{365}{5} \right\rfloor.$$

To this result we add the number of days where she gets at least 2 phone calls in a day because we double subtracted these days, which is

$$\left\lfloor \frac{365}{12} \right\rfloor + \left\lfloor \frac{365}{15} \right\rfloor + \left\lfloor \frac{365}{20} \right\rfloor.$$

We now subtract the number of days where she gets three phone calls, which is $\left\lfloor \frac{365}{60} \right\rfloor$. Therefore, our answer is

$$365 - \left(\left\lfloor \frac{365}{3} \right\rfloor + \left\lfloor \frac{365}{4} \right\rfloor + \left\lfloor \frac{365}{5} \right\rfloor \right) + \left(\left\lfloor \frac{365}{12} \right\rfloor + \left\lfloor \frac{365}{15} \right\rfloor + \left\lfloor \frac{365}{20} \right\rfloor \right) - \left\lfloor \frac{365}{60} \right\rfloor = 365 - 285 + 72 - 6 = \boxed{146}.$$

6.18 We start with complementary counting. After all, it's much easier to count the cases where some of these restraints are true, than when they aren't. PIE: Let's count the total number of cases where one of these is true:

- When Alice is with Bob: $2 \cdot 4! = 48$
- When Alice is with Carla: $2 \cdot 4! = 48$
- When Derek is with Eric: $2 \cdot 4! = 48$

Then, we count the cases where two of these are true.

- Alice is next to Carla, and Alice is also next to Bob. There are two ways to rearrange Alice, Bob, and Carla so that this is true: BAC and CAB. $2 \cdot 3! = 12$
- Alice is next to Carla, and Derek is also next to Eric. $2 \cdot 2 \cdot 3! = 24$
- Alice is next to Bob, and Derek is also next to Eric. $2 \cdot 2 \cdot 3! = 24$

Finally, we count the cases where all three of these are true: $2 \cdot 2 \cdot 2 = 8$. We add up the cases where one of these are true: $48 \cdot 3 = 144$. Subtract the cases where two of these are true: $144 - 60 = 84$, and add back the cases where three of these are true: $84 + 8 = 92$. Thus, our answer is $5! - 92 = \boxed{28}$.

6.19 Consider the three-digit arrangement, $\overline{aba}$. There are 10 choices for a and 10 choices for b (since it is possible for $a = b$), and so the probability of picking the palindrome is $\frac{10 \times 10}{10^3} = \frac{1}{10}$. Similarly, there is a $\frac{1}{26}$ probability of picking the three-letter palindrome. By the Principle of Inclusion-Exclusion, the total probability is

$$\frac{1}{26} + \frac{1}{10} - \frac{1}{260} = \frac{35}{260} = \frac{7}{52} \implies 7 + 52 = \boxed{059}$$

Essential Academy (June 2023)   **Introduction to AMC 8: Solutions Manual**

6.20 There are a total of 3^7 total assignments. To count the number of valid assignments, we subtract out the number of assignments that map to only one or two mentors. Mapping to one mentor, there are 3 assignments (all students are assigned to that mentor). For two mentors, there are $_3C_2 \cdot (2^7 - 2)$: $_3C_2$ ways to select the two mentors that get students and $2^7 - 2$ ways to perform the mappings. The reason we subtracted two is to subtract out the case where all three students select the same mentor, which we already accounted for. Thus, the final answer is $3^7 - (3 \cdot (2^7 - 2) + 3) = 3^7 - 3*2^7 + 3 = \boxed{1806}$.

6.21 The given states that there are 168 prime numbers less than 1000, which is a fact we must somehow utilize. Since there seems to be no easy way to directly calculate the number of "prime-looking" numbers, we can apply complementary counting. We can split the numbers from 1 to 1000 into several groups: $\{1\}$, $\{$numbers divisible by $2 = S_2\}$, $\{$numbers divisible by $3 = S_3\}$, $\{$numbers divisible by $5 = S_5\}$, $\{$primes not including $2, 3, 5\}$, $\{$prime-looking$\}$. Hence, the number of prime-looking numbers is $1000 - (168 - 3) - 1 - |S_2 \cup S_3 \cup S_5|$ (note that $2, 3, 5$ are primes).

We can calculate $S_2 \cup S_3 \cup S_5$ using the Principle of Inclusion-Exclusion: (the values of $|S_2|$... and their intersections can be found quite easily)

$$|S_2 \cup S_3 \cup S_5| = |S_2| + |S_3| + |S_5| - |S_2 \cap S_3| - |S_3 \cap S_5| - |S_2 \cap S_5| + |S_2 \cap S_3 \cap S_5|$$

$$= 500 + 333 + 200 - 166 - 66 - 100 + 33 = 734$$

Substituting, we find that our answer is $1000 - 165 - 1 - 734 = \boxed{100}$.

6.22 Let $A = \{2, 4, 6, 8, 10, 12\}$, $B = \{3, 6, 9, 12\}$, $C = \{2, 3, 5, 7, 11\}$, $U = \{1, 2, 3, \ldots, 12\}$. The total number of subsets of U is 2^{12}. Now let's count the number of subsets $S \subseteq U$ that have no element in common with at least one of A, B, C. We use Principle of Exclusion and Inclusion.

- Number of S such that $S \cap A = \emptyset$: $2^{12-6} = 2^6 = 64$
- Number of S such that $S \cap B = \emptyset$: $2^{12-4} = 2^8 = 256$
- Number of S such that $S \cap C = \emptyset$: $2^{12-5} = 2^7 = 128$

Now counting for the number of S that do not have common elements with 2 or more of our sets,

- Number of S such that $S \cap (A \cup B) = \emptyset$: $2^{12-8} = 2^4 = 16$
- Number of S such that $S \cap (B \cup C) = \emptyset$: $2^{12-8} = 2^4 = 16$
- Number of S such that $S \cap (C \cup A) = \emptyset$: $2^{12-9} = 2^3 = 8$

Now, finally, counting the number of S that do not have common elements with all 3 of the sets gives us: $2^{12-11} = 2$. Therefore, our final answer is

$$2^{12} - (2^6 + 2^8 + 2^7) + (2^4 + 2^4 + 2^3) - 2 = \boxed{3686}.$$

§7 Stars and Bars

7.1 $\boxed{45}$.

7.2 In order to distribute amongst 3 people, we need 2 bars that divide up our 8 pencils. Therefore, we just need to compute $_{10}C_8 = \boxed{45}$.

7.4 $\boxed{84}$.

7.5 $\boxed{120}$.

7.6 We can start off by giving each basket a banana and a peach to remove the constraint of "at least 1 of any fruit." Then, we just need to distribute 8 peaches and 2 bananas. The bananas can be split either $2, 0, 0$ or $1, 1, 0$ form.

Case 1: Basket 1, 2, or 3 receives 2 bananas.
Without loss of generality let's say basket 1 receives 2 bananas. Then, give 3 peaches to basket 1 to remove the constraint that we need more peaches than bananas. Now, we can freely distribute the remaining 5 peaches using stars and bars, which gives $_7C_2 = 21$. Multiplying by 3 to account for our initial assumption, this case gives $3 \cdot 21 = 63$.

Case 2: Basket 1, 2, or 3 does not receive 2 bananas.
In this case, two of the basket will receive 1 banana each. We have $_3C_2 = 3$ choices to assign these bananas. To remove the constraint that we need more peaches and bananas, give 2 peaches each to the baskets that received bananas. Now, we can freely distribute the remaining 4 peaches using stars and bars, which gives $_6C_2 = 15$. This case therefore gives $3 \cdot 15 = 45$.

Adding up the cases we get an answer of $\boxed{108}$.

Remark. Our problem is part of the "Stars and Bars" section, so assume the baskets are distinguishable.

7.7 By Stars and Bars, $_7C_5 = \boxed{21}$.

7.8 It is given that it is possible to select at least 6 of each. Therefore, we can make a bijection to the number of ways to divide the six choices into three categories, since it is assumed that their order is unimportant. Using the ball and urns/sticks and stones/stars and bars formula, the number of ways to do this is $_8C_2 = \boxed{28}$.

7.9 We use stars and bars. Let Alice get k apples, let Becky get r apples, let Chris get y apples.

$$\implies k + r + y = 24$$

We can manipulate this into an equation which can be solved using stars and bars. All of them get at least 2 apples, so we can subtract 2 from k, 2 from r, and 2 from y.

$$\implies (k - 2) + (r - 2) + (y - 2) = 18$$

Let $k' = k - 2$, let $r' = r - 2$, let $y' = y - 2$.

$$k' + r' + y' = 18$$

Essential Academy (June 2023)　　　　　　　　　Introduction to AMC 8: Solutions Manual

We can allow either of them to equal to 0; hence, this can be solved by stars and bars. By Stars and Bars, our answer is just $_{18+3-1}C_{3-1} = {_{20}}C_2 = \boxed{190}$.

7.10 There are 4 stars (items to be purchased), and five bars (to divide the six categories). Therefore, a total of 9 objects. So it's $_9C_4 = \boxed{126}$, just like the previous example.

7.11 Start by getting one rose of each color, so now you need to pick 9 more. There are 3 categories, therefore only need 2 bars. 9 stars and 2 bars equals 11 total objects. So it's $_{11}C_2 = \boxed{55}$.

7.12 We have 10 stars and 2 bars to distribute among our 3 categories. Therefore, our answer is $_{12}C_2 = \boxed{66}$.

7.13 The total possible number of outcomes is 6^4. To count the number of desirable outcomes, interpret the problem using stars and bars. The stars would be the 9 points we have to distribute and 3 bars would divide those points among the four dice. However, since the minimum number in a die is 1 we can start off by giving 1 point to each dice. After that, our constraints have been eliminated, so we have 5 stars to distribute using 3 bars, which tells us the number of desired outcomes is $_7C_3 = 35$. Therefore, our answer is $\frac{35}{9^4} = \boxed{\dfrac{35}{1296}}$.

18

§8 Geometric Counting

8.1 $\boxed{56}$.

8.2 $\boxed{30}$.

8.4 $\boxed{90}$.

8.6 Let's first count the number of rectangles. By **Lemma 8.5**, the number of rectangles is $_6C_2 \times {}_5C_2 = 150$. The number of squares is
$$5 \cdot 4 + 4 \cdot 3 + 3 \cdot 2 + 2 \cdot 1 = 40$$
Therefore, our desired count is $150 - 40 = \boxed{110}$.

8.7 A square of side length $\sqrt{5}$ fits inside a box of 3×3. There are 9 3×3 boxes in total and 2 configurations of inner $\sqrt{5}$ squares possible in a box of 3×3. Therefore, our answer is $9 \cdot 2 = \boxed{18}$.

8.8 $\boxed{252}$.

8.9 The total number of ways to go from $(0,0)$ to $(5,6)$ is $_{11}C_5$. From that total, let's subtract the number of ways to go from $(0,0)$ to $(1,2)$ to $(3,4)$ to $(5,5)$ to $(5,6)$ which is
$$_3C_1 \cdot {}_4C_2 \cdot {}_3C_1 \cdot 1 = 54.$$
Therefore, our answer is $\boxed{408}$.

8.10 Let increasing x and y represent moving right and upwards, respectively. Once Sally moves right she cannot move left again. Now consider Sally's movement within each column (each fixed x). At any given y-coordinate in that column, she can either travel right or travel vertically and then travel right. Since her path cannot overlap itself, for any y_1, y_2-coordinates in each column, there is exactly one way to travel from (x, y_1) to (x, y_2). Thus, for each of the first 4 columns Sally has 5 choices of y-coordinate on which she will travel right - once she reaches the column $x = 4$, there is only one way to travel up to $(4, 4)$. Thus, our answer is $5^4 = \boxed{625}$.

8.11 On any white square, we may choose to go left or right, as long as we do not cross over the border of the board. Call the moves L and R respectively. Every single legal path consists of 4 $R's$ and 3 $L's$, so now all we have to find is the number of ways to order $4R's$ and $3L's$ in any way, which is $_7C_3 = 35$. However, we originally promised that we will not go over the border, and now we have to subtract the paths that do go over the border. The paths that go over the border are any paths that start with RRR (1 path), RR (5 paths) and $LRRR$ (1 path) so our final number of paths is $35 - 7 = \boxed{28}$.

8.12 $_5C_2 \cdot {}_7C_2 = \boxed{210}$.

8.13 $_{10}C_5 = \boxed{252}$.

8.14 Using combinations, we get that the number of ways to get from Samantha's house to City Park is $_3C_1 = \frac{3!}{1!2!} = 3$, and the number of ways to get from City Park to school is $_4C_2 = \frac{4!}{2!2!} = \frac{4 \cdot 3}{2} = 6$. Since there's one way to go through City Park (just walking straight through), the number of different ways to go from Samantha's house to City Park to school $3 \cdot 6 = \boxed{18}$.

8.15 Note that the only way for Leanne to get to the point $(20, 20)$ is to go from (x, y) to $(x+1, y)$ 20 times and to $(x, y+1)$ 20 times, and the number of ways to do this is the number of orders she can do these two kinds of 20 moves in, or $_{40}C_{20}$. Similarly, if to get to some point (a, b) Jing Jing has to go from (x, y) to $(x-2, y+5)$ r times and to $(x+3, y-1)$ s times, then the number of ways to get to this point is the number of ways to order these two kinds of $r+s$ moves, or $_{r+s}C_r$. Thus, we want

$$_{r+s}C_r = {}_{40}C_{20}.$$

We have that $(a, b) = (0 + r \cdot (-2) + s \cdot 3, 0 + r \cdot 5 + s \cdot (-1)) = (20, 80)$. This means that $a + b = \boxed{100}$.

We now show that $(r, s) = (20, 20)$ is minimal. In other words, we show that there is no other solution (r, s) such that $3r + 2s \leq 100$. First, note that if $r + s > 40$, then $r < 20$. This means that $_{r+s}C_r$ would contain a factor of 41, which is not okay.

Thus, we need $r + s < 40$ to find another valid solution. But then we need $r > 20$. If $r \geq 23$, then we don't have enough factors of 23 in our coefficient. Thus, we only need to check cases where $r + s < 40, r \leq 23$, and $3r + 2s \leq 100$. Enumerating through all the solutions gives solutions that aren't equal to $_{40}C_{20}$, so $(r, s) = (20, 20)$ is the smallest solution.

§9 Recursion

9.1 $\boxed{13}$.

9.2 $\boxed{12}$.

9.3 Denote P_n to be the probability that the cricket would return back to the first point after n hops. Then, we get the recursive formula
$$P_n = \frac{1}{3}(1 - P_{n-1})$$
because if the leaf is not on the target leaf, then there is a $\frac{1}{3}$ probability that it will make it back.

With this formula and the fact that $P_1 = 0$ (After one hop, the cricket can never be back to the target leaf.), we have
$$P_2 = \frac{1}{3}, P_3 = \frac{2}{9}, P_4 = \frac{7}{27},$$
so our answer is $\boxed{\dfrac{7}{27}}$.

9.4 Let a_n be the number of ways to walk a n-stair staircase. Consider the first step. You can walk either 1, 2, or 3 steps, leaving $n-1$, $n-2$, $n-3$ steps respectively. Thus, we have $a_n = a_{n-1} + a_{n-2} + a_{n-3}$. Clearly, $a_0 = 1$, $a_1 = 1$, and $a_2 = 2$, so $a_3 = 4$, $a_4 = 7$, $a_5 = 13$, and $a_6 = \boxed{24}$.

9.5 Let S_n denote the number of spacy subsets of $\{1, 2, ...n\}$. We have $S_0 = 1, S_1 = 2, S_2 = 3$.

The spacy subsets of S_{n+1} can be divided into two groups:

- A = those not containing $n+1$. Clearly $|A| = S_n$.

- B = those containing $n+1$. We have $|B| = S_{n-2}$, since removing $n+1$ from any set in B produces a spacy set with all elements at most equal to $n-2$, and each such spacy set can be constructed from exactly one spacy set in B.

Hence, $S_{n+1} = S_n + S_{n-2}$. From this recursion, we find that the answer is $\boxed{129}$.

9.6 We proceed recursively. Suppose we can build T_m towers using blocks of size $1, 2, \ldots, m$. How many towers can we build using blocks of size $1, 2, \ldots, m, m+1$? If we remove the block of size $m+1$ from such a tower (keeping all other blocks in order), we get a valid tower using blocks $1, 2, \ldots, m$. Given a tower using blocks $1, 2, \ldots, m$ (with $m \geq 2$), we can insert the block of size $m+1$ in exactly 3 places: at the beginning, immediately following the block of size $m-1$ or immediately following the block of size m. Thus, there are 3 times as many towers using blocks of size $1, 2, \ldots, m, m+1$ as there are towers using only $1, 2, \ldots, m$. There are 2 towers which use blocks $1, 2$, so there are $2 \cdot 3^6 = 1458$ towers using blocks $1, 2, \ldots, 8$, so the answer is $\boxed{458}$.

(Note that we cannot say, "there is one tower using the block 1, so there are 3^7 towers using the blocks $1, 2, \ldots, 8$." The reason this fails is that our recursion only worked when $m \geq 2$: when $m = 1$, there are only 2 places to insert a block of size $m + 1 = 2$, at the beginning or at the end, rather than the 3 places we have at later stages. Also, note that this method generalizes directly to seeking the number of towers where we change the second rule to read, "The cube immediately on top of a cube with edge-length k must have edge-length at most $k+n$," where n can be any fixed integer.)

Essential Academy (June 2023)   **Introduction to AMC 8: Solutions Manual**

9.7 $\frac{13}{52} = \boxed{\frac{1}{4}}$.

9.8
$$\frac{26}{52} \cdot \frac{25}{51} \cdot \frac{24}{50} = \boxed{\frac{1}{221}}$$

9.9 Given you've selected the first card, the probability of picking a card of another suit is $\frac{39}{51} = \boxed{\frac{13}{17}}$.

9.10 There are three face cards (Jack, Queen, and King) for each suit. Therefore, our probability is
$$\frac{12}{52} \cdot \frac{11}{51} = \boxed{\frac{11}{221}}$$

9.11 The first card can be any of 4 suits but then for the next 3 card to be the same suit. This gives a probability of
$$\frac{12}{51} \cdot \frac{11}{50} \cdot \frac{10}{49} = \boxed{\frac{44}{4165}}$$

9.12 Simplify our steps:

- The probability of drawing a red Ace as the top card is $\frac{2}{52}$, which simplifies to $\frac{1}{26}$.
- The probability of drawing a spade as the second card is $\frac{13}{51}$.

Combining these probabilities gives $\boxed{\frac{13}{1326}}$.

9.13 We can split into two cases.

- If top card is Ace of Spades: $\frac{1}{52} \cdot \frac{12}{51}$
- If top card is an ace that is not a space: $\frac{3}{52} \cdot \frac{13}{51}$

They sum up to a total of $\boxed{\frac{1}{52}}$

9.14 Split cases based on whether the first card selected was an ace of spades ($\frac{1}{52}$), non-spade ace ($\frac{3}{52}$), or non-ace spade ($\frac{12}{52}$). Then, we have
$$\frac{1}{52} \cdot \frac{15}{51} + \frac{3}{52} \cdot \frac{13}{51} + \frac{12}{52} \cdot \frac{4}{51} = \boxed{\frac{1}{26}}.$$

9.15 For each suit, there are 13 ways to draw consecutive cards. These are: $\{A, 2\}, \{2, 3\}, \{3, 4\}, \{4, 5\}, \{5, 6\}, \{6, 7\}, \{7, 8\}, \{8, 9\}, \{9, 10\}, \{10, J\}, \{J, Q\}, \{Q, K\}$, and $\{K, A\}$. Therefore, given that we've drawn a card, (ex. 4 of spades) we have 2 choices for the second card (3 or 5 of spades). Therefore, our answer is $\frac{2}{52} = \boxed{\frac{1}{26}}$.

9.16 The previous outcomes of the three flips (all heads) do not affect the probability of the next flip. Each flip of the fair coin is an independent event, and the outcome of one flip does not influence the outcome of another. Therefore, the probability of getting heads on the next flip is $\boxed{\frac{1}{2}}$.

9.17 The probability of rolling a 5 on the red die is $\frac{1}{6}$ and the probability of rolling an even number on the green die is $\frac{1}{2}$. Hence, our answer is $\boxed{\dfrac{1}{12}}$.

9.18 For each bin, the probability of selecting an odd number is $\frac{5}{9}$. Therefore, our final answer is $\left(\frac{5}{9}\right)^3 = \boxed{\dfrac{125}{729}}$.

9.19 The last three digits range from $000 \sim 999$, so there is a total of 1000 possible outcomes. The number of outcomes where the last three digits are the same is 10 : $000, 111, \ldots, 999$. Therefore, our answer is
$$\frac{10}{10^3} = \boxed{\dfrac{1}{100}}$$

9.20 There are in total 6^6 possible outcomes and $6!$ ways to rearrange numbers $1, 2, \ldots, 6$. Therefore, our desired probability is
$$\frac{6!}{6^6} = \boxed{\dfrac{5}{324}}$$

9.21 There are in total 36 possible outcomes. After a dot is added, the faces will range from $2 \sim 7$. In order for them to sum to 8 we can either have $(2,6), (3,5), (4,4), (5,3), (6,2)$. $\frac{5}{6^2} = \boxed{\dfrac{5}{36}}$.

9.22 The total number of outcomes is $6! = 720$, which is the number of ways to arrange the 6 students. Since there is only one favorable outcome (the specific order from oldest to youngest), the number of favorable outcomes is 1. Hence, the answer is $\boxed{\dfrac{1}{720}}$.

9.23 We either want both numbers to be positive or negative. For each spinner, the probability that number is positive is $\frac{1}{2}$ Therefore, our answer is
$$\frac{1}{2} \cdot \frac{1}{2} + \frac{1}{2} \cdot \frac{1}{2} = \boxed{\dfrac{1}{4}}.$$

9.24 Let's split cases based on whether the sum of the first two digits is odd or even. If the sum of the first two digits is odd, then we need to select our third number to have a units digit that's even. This has a probability of $\frac{5}{10} = \frac{1}{2}$. If the sum of the first two digits is even, then we need to select whether our third number to have a units digit that's odd. This has a probability of $\frac{5}{10} = \frac{1}{2}$. Hence, for any choice for the first two digits, the probability the sum of the three units digits is even is $\boxed{\dfrac{1}{2}}$.

Algebra

§10 Ratio and Percentage

10.1 Given the rate, 2 of $12 \cdot 64$ calculators are defective. Therefore, in total, the number of defective calculators is
$$\frac{2}{12 \cdot 64} \cdot (64 \cdot 1444) = \boxed{24}$$

10.2
$$\frac{1}{2}\left(\frac{4}{10} \cdot \frac{1}{10} + \frac{2}{10} \cdot \frac{4}{10}\right) = \boxed{0.06}$$

10.3 Call the number of marbles in each jar x (because the problem specifies that they each contain the same number). Thus, $\frac{x}{10}$ is the number of green marbles in Jar 1, and $\frac{x}{9}$ is the number of green marbles in Jar 2. Since $\frac{x}{9} + \frac{x}{10} = \frac{19x}{90}$, we have $\frac{19x}{90} = 95$, so there are $x = 450$ marbles in each jar.

Because $\frac{9x}{10}$ is the number of blue marbles in Jar 1, and $\frac{8x}{9}$ is the number of blue marbles in Jar 2, there are $\frac{9x}{10} - \frac{8x}{9} = \frac{x}{90} = 5$ more marbles in Jar 1 than Jar 2. This means the answer is $\boxed{5}$.

10.4
$$\frac{5}{7}\left(\frac{2}{7} \cdot \frac{1}{4} + \frac{2}{7} \cdot \frac{3}{4} + \frac{3}{7} \cdot \frac{1}{2}\right) \cdot 994 = \boxed{355}$$

10.6 $\boxed{10}$.

10.8 Alex can solve 60 equations in 30 minutes, and Bob can solve 60 equations in 6 minutes. Therefore, by **Remark 10.7**, our answer is
$$\frac{2}{\frac{1}{30} + \frac{1}{6}} = \boxed{10 \text{ minutes}}$$

10.9 Let a, b be the rates of experienced workers and new workers, respectively, in houses per month. This gives
$$1 = 6(12a + 6b), \quad 1 = 9(6a + 12b) \implies a = 4b \implies a = \frac{1}{81}, b = \frac{1}{324}$$
Therefore, if there are 9 experienced workers and 9 new workers, $9a + 9b = \frac{1}{9} + \frac{1}{36} = \frac{5}{36}$, so it would take $\frac{36}{5} = \boxed{7.2}$ months to build a house.

10.10 Einstein works at a rate of $\frac{1}{36} = \frac{4}{144}$ rocket per hour while Oppenheimer works at a rate of $\frac{1}{48} = \frac{3}{144}$ rocket per hour. Working together they work at a rate of $\frac{1}{36} + \frac{1}{48} = \frac{7}{144}$ rocket per hour. For free, they can create $\frac{42}{144}$ rocket. So without extra pay, they would be able to complete it in 4 working days.

Now can we reduce this number to 3 with extra pay? $144 - 42 \cdot 3 = 18$. Paying Einstein $300 and Oppenheimer $200 would finish the job within 3 workdays. ($18 = 4 \cdot 3 + 3 \cdot 2$).

Now can we reduce this number to 2? $144 - 42 \cdot 2 = 60$. Paying Einstein $1000 wouldn't be able to get the job done as he would only be able to complete an extra $\frac{40}{144}$ of a rocket, but we need $\frac{60}{144}$.

Hence, the answer is $\boxed{3}$.

10.11 $180 \times \frac{4}{3+3+4} = \boxed{72}$

10.12 The number of girls in Colfax Middle School is $270 \cdot \frac{4}{9} = 120$.

The number of girls in Winthrop Middle School is $180 \cdot \frac{5}{9} = 100$.

The total number of students attending the dance is 450 and the total number of girls is 220. Therefore the fraction of the girls at the dance is $\frac{220}{450} = \boxed{\frac{22}{45}}$.

10.13 First, by giving 20% to Pedro, Gilda is left with 80% of her total marbles. By giving 10% of what is left to Ebony, Gilda is left with $80\% - (80\% \cdot \frac{1}{10}) = 72\%$ of her original marbles. Finally, Gilda gives 25% of what is now left, which makes her $72\% - (72\% \cdot \frac{25}{100}) = \boxed{54\%}$.

10.14 If we say the total number of marbles is x marbles, the numbers of each color of marbles are $\frac{x}{3}$ blue, $\frac{x}{4}$ red, and six green marbles. Therefore the number of yellow marbles is

$$x - \left(\frac{x}{3} + \frac{x}{4} + 6\right) = \frac{5x}{12} + 6$$

. Since the number of marbles should be integer, the smallest number of yellow marbles Marcy can have is when $x = 12$, so $5 + 6 = \boxed{11}$.

10.15 Using the ratio of 8th-graders to 6th-graders, we can say there are $5a$ 8th-graders and $3a$ 6th-graders. Again using the ratio for 8th-graders to 7th-graders, since the number of 8th-graders is $5a$, the number of 7th graders is $\frac{25a}{8}$. The total number of students is $8a + \frac{25a}{8}$. Since the number of students should be integer, the smallest number of students possible is when $a = 8$, which is $64 + 25 = \boxed{89}$.

10.16 Since the numbers of the first half of the assignment and the second half of the assignment are equal, Chloe's answer for the second half is 96%. Since Zoe had the same answer in the second half, Zoe's overall percentage of correct answers is $\frac{90+96}{2} = \boxed{93\%}$.

10.17
$$15\%x = 20\%y \iff y = 75\%x$$
$\therefore \boxed{75\%}$.

10.18 The number of students who didn't go to drama practice is $4 \cdot 6 = 24$. Since 24 is $\frac{2}{3}$ of the total number of students, $24 \cdot \frac{3}{2} = \boxed{36}$.

10.19 We will solve this problem by converting all the variables using z. Since the ratio of z to x is $1:6$, $x = 6z$. Then since the ratio of w to x is $4:3$, w is $6z$ times $\frac{4}{3}$, which is $8z$. Then, since the ratio of y to z is $3:2$, y equals $\frac{3}{2}z$. Therefore the ratio of w to y is the ratio of $8z$ to $\frac{3}{2}z$ is

$$\frac{8}{\frac{3}{2}} = \boxed{\frac{16}{3}}$$

10.20 If we say the shirt was increased by a percent, the ratio of original price to the final price is

$$100 : \frac{100+a}{\frac{100-a}{100}} = 100 : 84$$

. Using the ratio we get $8400 = 10000 - a^2$. Therefore the price of a shirt was increased and decreased by $\boxed{40\%}$.

10.21 Rate = Work/Time. Therefore, the ratio of Andre's rate of doing work is

$$\frac{\frac{5}{6}}{\frac{3}{4}} : 1 = \frac{10}{9} : 1 = \boxed{\frac{10}{9}}$$

10.22 To equalize all the units 4.2 megabyte = 4.2×8000 kilobits = 33600 kilobits. Using the rate of download speed,
$$56 : 1 = 33600 : x \text{ seconds}$$
$\therefore x = 600$ seconds Converting the seconds to minutes, it takes approximately $\boxed{10}$ minutes to download the song.

10.23 Call the number of marbles in each jar x (because the problem specifies that they each contain the same number). Thus, $\frac{x}{10}$ is the number of green marbles in Jar 1, and $\frac{x}{9}$ is the number of green marbles in Jar 2. Since $\frac{x}{9} + \frac{x}{10} = \frac{19x}{90}$, we have $\frac{19x}{90} = 95$, so there are $x = 450$ marbles in each jar.

Because $\frac{9x}{10}$ is the number of blue marbles in Jar 1, and $\frac{8x}{9}$ is the number of blue marbles in Jar 2, there are $\frac{9x}{10} - \frac{8x}{9} = \frac{x}{90} = 5$ more marbles in Jar 1 than Jar 2. This means the answer is $\boxed{5}$.

10.24 Without loss of generality, let total weight of grapes be 10 kg. Then weight of water= 8 kg and weight of pulp= 2 kg. If w is the weight of water after evaporation, $2 + w = 4$ gives $w = 2$. Hence, the amount of water loss of $8 - 2 = 6$ kg. Remaining are 2 kg of water, so $2/4 = \boxed{50\%}$ weight of water remains as a percentage of grapes.

10.25 Let's express the weights as decimal numbers rather than percentages: $20\% = 0.20$ to be applied to the lowest pre-final exam score, which is 65%; $25\% = 0.25$ to be applied to the other pre-final exam scores, 80% and 92%. The sum of the weights must be $100\% = 1.00$, so the final exam must have a weight of $1.00 - (0.20 + 0.25 + 0.25) = 0.30$. We are to assume the maximum possible score of 100% for the final exam. The weighted average is $0.20 \times 65 + 0.25 \times 80 + 0.25 \times 92 + 0.30 \times 100 = 13 + 20 + 23 + 30 = \boxed{86\%}$

§11 Linear Function and Quadratic Function

11.2 $x = 9, y = 13, z = 5$.

11.3 Summing the first two equations gives $2y = 12 \Longrightarrow y = 6$. Similarly, summing the second and third equation gives $z = 8$, and summing the first and third equation gives $x = 7$. Therefore, $x^2 + y^2 + z^2 = 6^2 + 7^2 + 8^2 = \boxed{149}$.

11.4 By the previous problem, $y = 6$ and $x = 7$. Therefore, $y^2 - x^2 = \boxed{-13}$.

11.5 $25 = x^2y + xy^2 = xy(x+y)$. As $xy = 4$, $x + y = \frac{25}{4}$. Therefore,

$$x^3y + x^2y^2 + xy^3 = xy(x^2 + y^2 + xy)$$
$$= xy((x+y)^2 - xy)$$
$$= 4\left(\frac{625}{16} - 4\right)$$
$$= \boxed{\frac{561}{4}}$$

11.6 Let's say Orangey initially drank $8a$ ounces of orange juice. Then, after filling in apple juice, in the glass would have been $8 - 8a$ orange juice and $8a$ apple juice. Orangey would then take $\frac{2}{3}(8 - 8a)$ orange juice and $\frac{2}{3}(8a)$ apple juice. After filling in water, remaining in the mixture would be $\frac{1}{3}(8 - 8a)$ orange juice, $\frac{1}{3}(8a)$ apple juice, $\frac{16}{3}$ water. After Orangey finishes the drink, we can compare the total amount of orange juice with the total amount of apple juice.

$$(8 - 8a) + \frac{2}{3}(8 - 8a) + \frac{1}{3}(8a) = \text{Orange Juice} = 4 \cdot \text{Apple Juice} = 4(8a + \frac{2}{3}(8a) + \frac{1}{3}(8a))$$

$$\therefore 2(8 - 8a) = 4(16a) \Longrightarrow a = \frac{1}{5}.$$

In total, Orangey therefore drank $8a + 8 + 8 = \boxed{\frac{88}{5}}$ ounces of liquid.

11.7 There are $\frac{45}{100}(4) = \frac{9}{5}$ L of acid in Jar A. There are $\frac{48}{100}(5) = \frac{12}{5}$ L of acid in Jar B. And there are $\frac{k}{100}$ L of acid in Jar C. After transferring the solutions from jar C, there will be $4 + \frac{m}{n}$ L of solution in Jar A and $\frac{9}{5} + \frac{k}{100} \cdot \frac{m}{n}$ L of acid in Jar A.

$6 - \frac{m}{n}$ L of solution in Jar B and $\frac{12}{5} + \frac{k}{100} \cdot \left(1 - \frac{m}{n}\right) = \frac{12}{5} + \frac{k}{100} - \frac{mk}{100n}$ of acid in Jar B. Since the solutions are 50% acid, we can multiply the amount of acid for each jar by 2, then equate them to the amount of solution.

$$\frac{18}{5} + \frac{km}{50n} = 4 + \frac{m}{n}$$
$$\frac{24}{5} - \frac{km}{50n} + \frac{k}{50} = 6 - \frac{m}{n}$$

Add the equations to get

$$\frac{42}{5} + \frac{k}{50} = 10$$

Solving gives $k = 80$. If we substitute back in the original equation we get $\frac{m}{n} = \frac{2}{3}$ so $3m = 2n$. Since m and n are relatively prime, $m = 2$ and $n = 3$. Thus $k + m + n = 80 + 2 + 3 = \boxed{085}$.

Essential Academy (June 2023)   Introduction to AMC 8: Solutions Manual

11.8 We can compare each of the scores with the average of 81: $76 \to -5$, $94 \to +13$, $87 \to +6$, $100 \to +19$;

So the last one has to be -33 (since all the differences have to sum to 0), which corresponds to $81 - 33 = \boxed{48}$.

11.9 Analyze the facts: 6 are blue and green, meaning $b + g = 6$; 8 are red and blue, meaning $r + b = 8$; 4 are red and green, meaning $r + g = 4$. Then we need to add these three equations: $b + g + r + b + r + g = 2(r + g + b) = 6 + 8 + 4 = 18$. It gives us all of the marbles are $r + g + b = 18/2 = 9$. So the answer is $\boxed{9}$.

11.10 $abcdef = (abc)(def) = 4 \cdot 18 = \boxed{72}$

11.11 We can start with the full score, 50, and subtract not only 2 points for each incorrect answer but also the 5 points we gave her credit for. This expression is equivalent to her score, 29. Let x be the number of questions she answers correctly. Then, we will represent the number incorrect by $10 - x$.

$$50 - 7(10 - x) = 29$$
$$50 - 70 + 7x = 29$$
$$7x = 49$$
$$x = \boxed{7}$$

11.12 $x + y = 37, y + z = 41, z + x = 44$ implies $x + y + z = (37 + 41 + 44)/2 = 61$. This gives $x = 20, y = 17, z = 24$. $\therefore xyz = \boxed{8160}$.

11.13 Let x be the number of pages in the book. After the first day, Hui had $\frac{4x}{5} - 12$ pages left to read. After the second, she had $\left(\frac{3}{4}\right)\left(\frac{4x}{5} - 12\right) - 15 = \frac{3x}{5} - 24$ left. After the third, she had $\left(\frac{2}{3}\right)\left(\frac{3x}{5} - 24\right) - 18 = \frac{2x}{5} - 34$ left. This is equivalent to 62.

$$\frac{2x}{5} - 34 = 62$$
$$2x - 170 = 310$$
$$2x = 480$$
$$x = \boxed{240}$$

11.14 Let n be the number of pre-district games. Therefore, we can write the percentage of total games won as a weighted average, namely $.45(n) + .75(8) = (n + 8)(.5)$. Solving this equation for n gives 40, but since the problem asked for all games, the answer is $n + 8 = 40 + 8 = \boxed{48}$.

11.15 Let the amount of 1 dollar socks be a, 3 dollar socks be b, and 4 dollar socks be c. We then know that $a + b + c = 12$ and $a + 3b + 4c = 24$. We can make $a + b + c = 12$ into $a = 12 - b - c$ and then plug that into the other equation, producing $12 - b - c + 3b + 4c = 24$ which simplifies to $2b + 3c = 12$. It's not hard to see $b = 3$ and $c = 2$. Now that we know b and c, we know that $a = 7$, meaning the number of 1 dollar socks Ralph bought is $\boxed{7}$.

11.16 Let us plug in $(5 \circ x) = 1$ into $3a - b$. Thus it would be $3(5) - x$. Now we have $2 * (15 - x) = 1$.

29

Plugging $2*(15-x)$ into $3a-b$, we have $6-15+x=1$. Solving for x we have
$$-9+x=1 \Longrightarrow x = \boxed{10}$$

11.17 Since there are an equal number of juniors and seniors on the debate team, suppose there are x juniors and x seniors. This number represents $25\% = \frac{1}{4}$ of the juniors and $10\% = \frac{1}{10}$ of the seniors, which tells us that there are $4x$ juniors and $10x$ seniors. There are 28 juniors and seniors in the program altogether, so we get
$$10x + 4x = 28 \Longrightarrow x = 2$$
Which means there are $4x = 8$ juniors on the debate team. $\boxed{8}$.

11.18 Let x be the number of 5-cent coins that Joe has. Therefore, he must have $(x+3)$ 10-cent coins and $(23-(x+3)-x)$ 25-cent coins. Since the total value of his collection is 320 cents, we can write
$$5x + 10(x+3) + 25(23-(x+3)-x) = 320$$
$$5x + 10x + 30 + 500 - 50x = 320$$
$$35x = 210$$
$$x = 6.$$
Joe has six 5-cent coins, nine 10-cent coins, and eight 25-cent coins. Thus, our answer is $8-6 = \boxed{2}$.

11.19 Assign a variable to the number of eggs Mia has, say m. Then, because we are given that Sofia has twice the number of eggs Mia has, Sofia has $2m$ eggs, and Pablo, having three times the number of eggs as Sofia, has $6m$ eggs.

For them to all have the same number of eggs, they must each have $\frac{m+2m+6m}{3} = 3m$ eggs. This means Pablo must give $2m$ eggs to Mia and m eggs to Sofia, so the answer is $\frac{m}{6m} = \boxed{\frac{1}{6}}$.

11.20 The units digit of a multiple of 10 will always be 0. We add a 0 whenever we multiply by 10. So, removing the units digit is equal to dividing by 10. Let the smaller number (the one we get after removing the units digit) be a. This means the bigger number would be $10a$. We know the sum is $10a + a = 11a$ so $11a = 17402$. So $a = 1582$. The difference is $10a - a = 9a$. So, the answer is $9(1582) = \boxed{14,238}$.

11.21 Let $x = a+1 = b+2 = c+3 = d+4 = a+b+c+d+5$. Since one of the sums involves $a, b, c,$ and d, it makes sense to consider $4x$. We have $4x = (a+1) + (b+2) + (c+3) + (d+4) = a+b+c+d+10 = 4(a+b+c+d) + 20$. Rearranging, we have $3(a+b+c+d) = -10$, so $a+b+c+d = \frac{-10}{3}$. Thus, our answer is $\boxed{-10/3}$.

11.22 Solving the system of equations, we get $a = \frac{1}{3}, b = \frac{4}{3}, c = -\frac{2}{3}$. Therefore, $741a + 825b + 639c = \boxed{\frac{1375}{9}}$.

11.23 Factorizing 4930 gives $2 \cdot 5 \cdot 17 \cdot 29$. We can see that the pair $2 \cdot 29 = 58$ and $5 \cdot 17 = 85$ work. Therefore, our answer is $58 + 85 = \boxed{143}$. **11.24** They meet 75 km away after point A. The first car travels 75 km and the second car travels $75(1 + \frac{x}{60})$. The time they travel are the same so,
$$\frac{75 \text{ km}}{60 \text{ km/h}} = \frac{75(1+\frac{x}{60}) \text{ km}}{75 \text{ km/h}} \Longrightarrow x = \boxed{15 \text{ minutes}}$$

11.25 Note that the sum of the first two numbers is $21 \cdot 2 = 42$, the sum of the middle two numbers is $26 \cdot 2 = 52$, and the sum of the last two numbers is $30 \cdot 2 = 60$.

It follows that the sum of the four numbers is $42 + 60 = 102$, so the sum of the first and last numbers is $102 - 52 = 50$. Therefore, the average of the first and last numbers is $50 \div 2 = \boxed{25}$.

11.26 Let A be the age of Ana and B be the age of Bonita. Then,
$$A - 1 = 5(B - 1), \quad A = B^2$$
Substituting the second equation into the first gives us
$$B^2 - 1 = 5(B - 1).$$
By using difference of squares and dividing, $B = 4$. Moreover, $A = B^2 = 16$. The answer is $16 - 4 = \boxed{12}$.

11.27 If there are x girls, then there are $x + 2$ boys. She gave each girl x jellybeans and each boy $x + 2$ jellybeans, for a total of $x^2 + (x+2)^2$ jellybeans. She gave away $400 - 6 = 394$ jellybeans.
$$x^2 + (x+2)^2 = 394$$
$$x^2 + x^2 + 4x + 4 = 394$$
$$2x^2 + 4x = 390$$
$$x^2 + 2x = 195$$

From here, we can see that $x = 13$ as $13^2 + 26 = 195$, so there are 13 girls, $13 + 2 = 15$ boys, and $13 + 15 = \boxed{28}$ students.

11.28 Say there are $12x$ sets of twins, $4x$ sets of triplets, and x sets of quadruplets. That's $12x \cdot 2 = 24x$ twins, $4x \cdot 3 = 12x$ triplets, and $x \cdot 4 = 4x$ quadruplets. A tenth of the babies are quadruplets and that's $\frac{1}{10}(1000) = \boxed{100}$.

11.29 Call the number of marbles in each jar x (because the problem specifies that they each contain the same number). Thus, $\frac{x}{10}$ is the number of green marbles in Jar 1, and $\frac{x}{9}$ is the number of green marbles in Jar 2. Since $\frac{x}{9} + \frac{x}{10} = \frac{19x}{90}$, we have $\frac{19x}{90} = 95$, so there are $x = 450$ marbles in each jar.

Because $\frac{9x}{10}$ is the number of blue marbles in Jar 1, and $\frac{8x}{9}$ is the number of blue marbles in Jar 2, there are $\frac{9x}{10} - \frac{8x}{9} = \frac{x}{90} = 5$ more marbles in Jar 1 than Jar 2. This means the answer is $\boxed{5}$.

11.30 With some trial and error, we can observe that $1, 9, 16$ work. Therefore, the answer is $1^2 + 9^2 + 16^2 = \boxed{338}$.

11.31 Let the income amount be denoted by A. We know that $\frac{A(p+.25)}{100} = \frac{28000p}{100} + \frac{(p+2)(A-28000)}{100}$. We can now try to solve for A:
$$(p + .25)A = 28000p + Ap + 2A - 28000p - 56000$$
$$.25A = 2A - 56000$$
$$A = \boxed{32000}$$

11.32 Let Paula work at a rate of p, the two helpers work at a combined rate of h, and the time it takes to eat lunch be L, where p and h are in house/hours and L is in hours. Then the labor on Monday, Tuesday, and Wednesday can be represented by the three following equations:

$$(8 - L)(p + h) = 50$$
$$(6.2 - L)h = 24$$
$$(11.2 - L)p = 26$$

With three equations and three variables, we need to find the value of L. Adding the second and third equations together gives us $6.2h + 11.2p - L(p + h) = 50$. Subtracting the first equation from this new one gives us $-1.8h + 3.2p = 0$, so we get $h = \frac{16}{9}p$. Plugging into the second equation:

$$(6.2 - L)\frac{16}{9}p = 24$$
$$(6.2 - L)p = \frac{27}{2}$$

We can then subtract this from the third equation:

$$5p = 26 - \frac{27}{2}$$
$$p = \frac{5}{2}$$

Plugging p into our third equation gives:

$$L = \frac{4}{5}$$

Converting L from hours to minutes gives us $L = 48$ minutes, which is $\boxed{48}$.

§12 Speed, Distance, and Time

12.2
$$40 \text{ miles} \times \frac{1 \text{ hour}}{60 \text{ miles}} \times \frac{60 \text{ minutes}}{1 \text{ hour}} = \boxed{40 \text{ minutes}}$$

12.3 On the way forward, Ronaldo takes $\frac{2}{8} = \frac{1}{4}$ hour. On the way back, Ronaldo runs 8 miles per hour for $\frac{1}{8}$ hour and runs 4 miles per hour for $\frac{1}{4}$ hour. Therefore, Ronaldo runs at an average rate of
$$\frac{1}{4} \cdot \frac{8}{12} + \frac{1}{8} \cdot \frac{4}{12} = \boxed{\frac{5}{24}}$$

12.4 From the question
$$\frac{60}{x}, \frac{60}{x+5}, \frac{60}{x+10}, \frac{60}{x+15} \in \mathbb{N}$$
so $x = 5$ is the only one that works. Hence, the answer is
$$\frac{60}{5} + \frac{60}{10} + \frac{60}{15} + \frac{60}{20} = \boxed{25}$$

12.5 Since Zia will wait for the bus if the bus is at the previous stop, we can create an equation to solve for when the bus is at the previous stop. The bus travels $\frac{1}{3}$ of a stop per minute, and Zia travels $\frac{1}{5}$ of a stop per minute. Now we create the equation, $\frac{1}{3}m = \frac{1}{5}m + 3 - 1$ (the -1 accounts for us wanting to find when the bus reaches the stop before Zia's). Solving, we find that $m = 15$. Now Zia has to wait 2 minutes for the bus to reach her, so our answer is $15 + 2 = \boxed{17}$.

12.6 We know that in the same amount of time given, Ina will run twice the distance of Eve, and Paul would run quadruple the distance of Eve. Let's consider the time it takes for Paul to meet Eve: Paul would've run 4 times the distance of Eve, which we can denote as d. Thus, the distance between B and C is $4d + d = 5d$. In that given time, Ina would've run twice the distance d, or $2d$ units.

Now, when Paul starts running back towards A, the same amount of time would pass since he will meet Ina at his starting point. Thus, we know that he travels another $4d$ units and Ina travels another $2d$ units.

Therefore, drawing out the diagram, we find that $2d + 2d + 4d + d = 9d = 1800 \implies d = 200$, and distance between A and B is the distance Ina traveled, or $4d = 4(200) = \boxed{800}$.

12.8 $\boxed{\dfrac{2ab}{a+b}}$.

12.9
$$\frac{2 \cdot 2 \cdot 7}{2 + 7} = \boxed{\frac{28}{9}}$$

12.10 To calculate the average speed, simply evaluate the total distance over the total time. Let the number of additional miles he has to drive be x. Therefore, the total distance is $15 + x$ and the total time (in hours) is
$$\frac{15}{30} + \frac{x}{55} = \frac{1}{2} + \frac{x}{55}.$$

We can set up the following equation:
$$\frac{15+x}{\frac{1}{2}+\frac{x}{55}}=50.$$
Simplifying the equation, we get
$$15+x=25+\frac{10x}{11}.$$
Solving the equation yields $x=110$, so our answer is $\boxed{110}$.

12.11 Let $2d$ miles be the distance from the trailhead to the fire tower, where $d>0$. When Chantal meets Jean, the two have traveled for
$$\frac{d}{4}+\frac{d}{2}+\frac{d}{3}=d\left(\frac{1}{4}+\frac{1}{2}+\frac{1}{3}\right)=d\left(\frac{3}{12}+\frac{6}{12}+\frac{4}{12}\right)=\frac{13}{12}d$$
hours. At that point, Jean has traveled for d miles, so his average speed is $\frac{d}{\frac{13}{12}d}=\boxed{\frac{12}{13}}$ miles per hour.

12.12 Every 10 feet Bella goes, Ella goes 50 feet, which means a total of 60 feet. They need to travel that 60 feet $10560\div 60 = 176$ times to travel the entire 2 miles. SInce Bella goes 10 feet 176 times, this means that she travels a total of 1760 feet. And since she walks 2.5 feet each step, $1760 \div 2.5 = \boxed{704}$

12.13 Writing the equation gives us: $\frac{d}{v}=\frac{1}{3}$ and $\frac{d}{v+18}=\frac{1}{5}$. This gives $d=\frac{1}{5}v+3.6=\frac{1}{3}v$, which gives $v=27$, which then gives $d=\boxed{9}$.

12.14 WLOG, let Al's speed be 15 steps per second, so Bob's speed is 5 steps per second. Then, Al was on the escalator for $\frac{150}{15}=10$ seconds and Bob was on the escalator for $\frac{75}{5}=15$ seconds. Let r be the rate of the escalator, in steps per second. Then, the total amount of steps is $150-10r=75+15r$. Al is getting 10 seconds of resistance at rate r from the escalator, while Bob is getting 15 seconds of help at rate r. Solving for r, we hvae $r=3$ steps per second. Then, we can plug r into the previous equation or subtract/add it to Al/Bob's rate (respectively) then multiply by their respective time. Either way, we get $\boxed{120}$ and we are done.

12.15 Joe drives at 30 miles per hour for 2 hours. Joe also drives at 45 miles per hour for $\frac{1}{3}$ hour. Therefore, the average speed is
$$30\cdot\frac{2}{2+\frac{1}{3}}+45\cdot\frac{\frac{1}{3}}{2+\frac{1}{3}}=\boxed{\frac{225}{7}}$$

12.16 Jones catches up 800 m in 4 min and thus, at a rate of: $\frac{800\text{m}}{4\text{min}}=\frac{200\text{m}}{1\text{min}}$ which equals
$$\frac{200\text{m}}{\text{min}}\times\frac{1\text{ km}}{1000\text{m}}\times\frac{60\text{ min}}{1\text{h}}=12\frac{\text{km}}{\text{h}}.$$
Jones is going 50 km/h, so the car is going $(50-12)$ km/h $=\boxed{38}$ km/h.

12.17 Let the total distance be d. Then $d=35(t+1)$. Since 1 hour has passed, and he increased his speed by 15 miles per hour, and he had already traveled 35 miles, the new equation is $d-35=50(t-1-\frac{1}{2})=50(t-\frac{3}{2})$. Solving, $d=35t+35=50t-40$, $15t=75$, $t=5$, and
$$d=35(5+1)=35\cdot 6=210\implies\boxed{210}$$

12.18 Let the total number of steps be x, the speed of the escalator be e and the speed of Jacob and Alexander be $2v, v$, respectively (all in steps/sec). The time it takes Jacob to take 27 steps is the same as the time it takes Jacob and the escalator to take x steps. Therefore,
$$\frac{27}{2v} = \frac{x}{e+2v} = \frac{x-27}{e}$$
The time it takes Alexander to take 18 steps is the same as the time it takes Alexander and the escalator to take x steps. Therefore,
$$\frac{18}{v} = \frac{x}{e+v} = \frac{x-18}{e}$$
Combining these two equations gives
$$\frac{3}{4} \cdot \frac{x-18}{e} = \frac{x-27}{e} \implies x = \boxed{54}$$

12.19 Let x be the length of the ship. Then, in the time that Emily walks 210 steps, the ship moves $210 - x$ steps. Also, in the time that Emily walks 42 steps, the ship moves $x - 42$ steps. Since the ship and Emily both travel at some constant rate, $\frac{210}{210-x} = \frac{42}{x-42}$. Dividing both sides by 42 and cross multiplying, we get $5(x - 42) = 210 - x$, so $6x = 420$, and $x = \boxed{70}$.

12.20 We know that in the same amount of time given, Ina will run twice the distance of Eve, and Paul would run quadruple the distance of Eve. Let's consider the time it takes for Paul to meet Eve: Paul would've run 4 times the distance of Eve, which we can denote as d. Thus, the distance between B and C is $4d + d = 5d$. In that given time, Ina would've run twice the distance d, or $2d$ units.

Now, when Paul starts running back towards A, the same amount of time would pass since he will meet Ina at his starting point. Thus, we know that he travels another $4d$ units and Ina travels another $2d$ units.

Therefore, drawing out the diagram, we find that $2d + 2d + 4d + d = 9d = 1800 \implies d = 200$, and distance between A and B is the distance Ina traveled, or $4d = 4(200) = \boxed{800}$

12.21 We can set up a system of equations. Butch and Sundance spend their entire time either walking or riding. Let B be the number of miles that Butch walks, b be the number of miles that he rides, S be the number of miles that Sundance walks, and s be the number of miles that Sundance rides. Since Butch and Sundance travel the same distance, we get that
$$B + b = S + s$$
We van also find that to travel 1 mile, it takes Butch 15 minutes, Sundance 24 minutes, and Sparky 10 minutes. Since Butch and Sundance never stop, they must have traveled for the same amount of time, or
$$15B + 10b = 24S + 10s$$
Multiplying the first equation by 10 and subtracting, we get
$$5B = 14S$$
The smallest positive integral values of B and S that satisfy the equation are $B = 14$ and $S = 5$, so Butch walks for 14 miles and Sundance walks for 5, a total of 19 miles. Plugging these values into the second equation (given that Butch and Sundance must be riding when they are not walking), we find that they spent 260 minutes, for and answer of $260 + 19 = \boxed{279}$.

Essential Academy (June 2023)   Introduction to AMC 8: Solutions Manual

§13 Sequence and Series

13.6 Let a be the first term and d be the common difference. Then, we have equations

$$5a + 10d = 5a + (1+2+3+4)d = 65, \quad 10a + 45d = 10a + (1+2+3+\cdots+9)d = 255.$$

which implies

$$25d = 10a + 45d - 2(5a + 10d) = 255 - 130 = 125 \implies d = 5 \implies a = 3.$$

Therefore, the sum of the first 15 equals

$$15a + (1+2+3+\cdots+14)d = 15a + 105d = 15(3) + 105(5) = \boxed{570}$$

13.7 Given the information,

$$137 = a_1 + a_2 + \cdots + a_{98} = 98 \cdot \frac{a_1 + a_{98}}{2}$$

This implies that our desired sum equals

$$a_2 + a_4 + \cdots + a_{98} = 49 \cdot \frac{a_2 + a_{98}}{2} = 49 \cdot \left(\frac{a_1 + a_{98}}{2} + \frac{1}{2}\right) = \frac{137}{2} + \frac{49}{2} = \boxed{93}$$

13.12 Assume that the largest geometric number starts with a 9. We know that the common ratio must be a rational of the form $k/3$ for some integer k, because a whole number should be attained for the 3rd term as well. When $k = 1$, the number is 931. When $k = 2$, the number is 964. When $k = 3$, we get 999, but the integers must be distinct. By the same logic, the smallest geometric number is 124. The largest geometric number is 964 and the smallest is 124. Thus the difference is $964 - 124 = \boxed{840}$.

13.13 We should realize that the two equations are 100 terms apart, so by subtracting the two equations in a form like

$$(a_{101} - a_1) + (a_{102} - a_2) + \ldots + (a_{200} - a_{100}) = 200 - 100 = 100$$

we get the value of the common difference of every hundred terms one hundred times. So we have to divide the answer by one hundred to get $\frac{100}{100} = 1$, which is the common difference of every hundred terms. Then we have to simply divide the answer by hundred again to find the common difference between one term, therefore our answer is $\frac{1}{100} = \boxed{0.01}$

13.14 Let d be the common difference. Then 9, $9 + d + 2 = 11 + d$, $9 + 2d + 20 = 29 + 2d$ are the terms of the geometric progression. Since the middle term is the geometric mean of the other two terms, $(11 + d)^2 = 9(2d + 29) \implies d^2 + 4d - 140 = (d + 14)(d - 10) = 0$. The smallest possible value occurs when $d = -14$, and the third term is $2(-14) + 29 = 1 \implies \boxed{1}$.

13.15 After the adding of the odd numbers, the total of the sequence increases by $836 - 715 = 121 = 11^2$. Since the sum of the first n positive odd numbers is n^2, there must be 11 terms in the sequence, so the mean of the sequence is $\frac{715}{11} = 65$. Since the first, last, and middle terms are centered around the mean, our final answer is $65 \cdot 3 = \boxed{195}$

13.16 The sequence provided is an arithmetic sequence with first term 1 and common difference 4, so the 100th term of the sequence is $1 + 4(99) = \boxed{395}$.

13.17 Let x be that difference. Then $b = a + x, c = a + 2x, d = a + 3x, e = a + 4x$. This gives
$$a + b + c + d + e = 5a + 10x = 440 \Longrightarrow a + 2x = 88.$$
For e to be as large as possible, $a = 0$. So, $x = 44$ and $e = 4 \cdot 44 = \boxed{176}$.

13.18 Let n be the 13th consecutive even integer that's being added up. By doing this, we can see that the sum of all 25 even numbers will simplify to $25n$ since $(n-2k) + \cdots + (n-4) + (n-2) + (n) + (n+2) + (n+4) + \cdots + (n+2k) = 25n$. Now, $25n = 10000 \to n = 400$. Remembering that this is the 13th integer, we wish to find the 25th, which is $400 + 2(25-13) = \boxed{424}$.

13.19 The sum of the first n odd numbers is n^2. As in our case $n^2 = 100$, we have $n = \boxed{10}$.

§14 Mean, Median, and Mode

14.2 $\boxed{16}$.

14.3 Suppose that S has n numbers other than the 68, and the sum of these numbers is s. We are given that
$$\frac{s+68}{n+1} = 56,$$
$$\frac{s}{n} = 55.$$
Clearing denominators, we have
$$s + 68 = 56n + 56,$$
$$s = 55n.$$
Subtracting the equations, we get $68 = n + 56$, from which $n = 12$. It follows that $s = 660$.

The sum of the twelve remaining numbers in S is 660. To maximize the largest number, we minimize the other eleven numbers: We can have eleven 1s and one $660 - 11 \cdot 1 = \boxed{649}$.

14.4 $\boxed{13}$.

14.5 As the unique mode is 8, there are at least two 8s. As the range is 8 and one of the numbers is 8, the largest one can be at most 16.

If the largest one is 16, then the smallest one is 8, and thus the mean is strictly larger than 8, which is a contradiction.

If we have 2 8's we can add find the numbers $4, 6, 7, 8, 8, 9, 10, 12$. This is a possible solution but has not reached the maximum.

If we have 4 8's we can find the numbers $6, 6, 6, 8, 8, 8, 8, 14$.

We can also see that they satisfy the need for the mode, median, and range to be 8. This means that the answer will be $\boxed{14}$.

14.6 Since 90 is 20 more than 70, and 80 is 10 more than 70, for 70 to be the average, the other number must be 30 less than 70, or $\boxed{40}$.

14.7 We first notice that the median name will be the $(19+1)/2 = 10^{\text{th}}$ name. The 10th name is $\boxed{4}$.

14.8 The mean, or average number of days is the total number of days divided by the total number of students. The total number of days is $1 \cdot 1 + 2 \cdot 3 + 3 \cdot 2 + 4 \cdot 6 + 5 \cdot 8 + 6 \cdot 3 + 7 \cdot 2 = 109$. The total number of students is $1 + 3 + 2 + 6 + 8 + 3 + 2 = 25$. Hence, $\frac{109}{25} = \boxed{4.36}$.

14.9 Listing the elements from least to greatest, we have $(5, 5, 6, 8, 106)$, we see that the median weight is 6 pounds. The average weight of the five kids is $\frac{5+5+6+8+106}{5} = \frac{130}{5} = 26$. Hence,
$$26 - 6 = \boxed{\text{average, by } 20}.$$

14.10 Putting the numbers in numerical order we get the list $0, 0, 1, 2, 3, 3, 3, 4$. The mode is 3. The median is $\frac{2+3}{2} = 2.5$. The average is $\frac{0+0+1+2+3+3+3+4}{8} = \frac{16}{8} = 2$. The sum of all three is $3 + 2.5 + 2 = \boxed{7.5}$

14.11 We know the unique mode must be 6, so the mean must be the same number 6. Let's imagine a scale. 6 exactly stands the mid-point of the scale. Numbers of $3, 4, 5$ represent the left side "weights" of the scale. Numbers of $6, 7, x$ represent the right side "weights" of the scale. On the left side, the difference of the three "weights" between 6 are $-3, -2, -1$, respectively. It gives us the total difference is -6. In order to allow the scale to keep balance, on the right side, the total difference must be $+6$. Because we have already known the difference of the right side "weights" between 6 is $0 + 1 = 1$, partially, so the difference between 6 and unknown x must be $+6 - 1 = +5$. It exactly gives us the answer: $x = 6 + 5 = \boxed{11}$

14.12 Since there is an odd number of terms, the median is the number in the middle, specifically, the third largest number is 18, and there are 2 numbers less than 18 and 2 numbers greater than 18. The sum of these integers is $5(15) = 75$, since the mean is 15. To make the largest possible number with a given sum, the other numbers must be as small as possible. The two numbers less than 18 must be positive and distinct, so the smallest possible numbers for these are 1 and 2. The number right after 18 also needs to be as small as possible, so it must be 19. This means that the remaining number, the maximum possible value for a number in the set, is $75 - 1 - 2 - 18 - 19 = 35, \boxed{35}$.

14.13 In order to maximize the median, we need to make the first half of the numbers as small as possible. Since there are 100 people, the median will be the average of the 50th and 51st largest amount of cans per person. To minimize the first 49, they would each have one can. Subtracting these 49 cans from the 252 cans gives us 203 cans left to divide among 51 people. Taking $\frac{203}{51}$ gives us 3 and a remainder of 50. Seeing this, the largest number of cans the 50th person could have is 3, which leaves 4 to the rest of the people. The average of 3 and 4 is 3.5. Thus our answer is $\boxed{3.5}$.

14.14 The sum of the reciprocals is $\frac{1}{1} + \frac{1}{2} + \frac{1}{4} = \frac{7}{4}$. Their average is $\frac{7}{12}$. Taking the reciprocal of this gives $\boxed{\frac{12}{7}}$.

14.15 Since there will be 9 elements after removal, and their mean is 6, we know their sum is 54. We also know that the sum of the set pre-removal is 66. Thus, the sum of the 2 elements removed is $66 - 54 = 12$. There are only $\boxed{5}$ subsets of 2 elements that sum to 12: $\{1, 11\}, \{2, 10\}, \{3, 9\}, \{4, 8\}, \{5, 7\}$.

14.16 First investigate the mean, median, and mode:

- The mean is $\frac{10+2+5+2+4+2+x}{7} = \frac{25+x}{7}$.

- Arranged in increasing order, the list is $2, 2, 2, 4, 5, 10$, so the median is either $2, 4$ or x depending upon the value of x.

- The mode is 2, since it appears three times.

We apply casework upon the median:

- If the median is 2 ($x \leq 2$), then the arithmetic progression must be constant.

- If the median is 4 ($x \geq 4$), because the mode is 2, the mean can either be $0, 3, 6$ to form an arithmetic progression. Solving for x yields $-25, -4, 17$ respectively, of which only 17 works because it is larger than 4.

39

Essential Academy (June 2023)   Introduction to AMC 8: Solutions Manual

- If the median is x ($2 \leq x \leq 4$), we must have the arithmetic progression $2, x, \frac{25+x}{7}$. Thus, we find that $2x = 2 + \frac{25+x}{7}$ so $x = 3$.

The answer is $3 + 17 = \boxed{20}$.

14.17 Let $S = \{a_1, a_2, a_3, \ldots, a_n\}$ with $a_1 < a_2 < a_3 < \ldots < a_n$. We are given the following:

$$\begin{cases} \sum_{i=1}^{n-1} a_i = 32(n-1) = 32n - 32, \\ \sum_{i=2}^{n} a_i = 40(n-1) = 40n - 40, \\ \sum_{i=2}^{n-1} a_i = 35(n-2) = 35n - 70, \\ a_n - a_1 = 72 \implies a_1 + 72 = a_n. \end{cases}$$

Subtracting the third equation from the sum of the first two, we find that

$$\sum_{i=1}^{n} a_i = (32n - 32) + (40n - 40) - (35n - 70) = 37n - 2.$$

Furthermore, from the fourth equation, we have

$$\sum_{i=2}^{n} a_i - \sum_{i=1}^{n-1} a_i = \left[(a_1 + 72) + \sum_{i=2}^{n-1} a_i\right] - \left[(a_1) + \sum_{i=2}^{n-1} a_i\right] = (40n - 40) - (32n - 32).$$

Combining like terms and simplifying, we have

$$72 = 8n - 8 \implies 8n = 80 \implies n = 10.$$

Thus, the sum of the elements in S is $37 \cdot 10 - 2 = 368$, and since there are 10 elements in S, the average of the elements in S is $\frac{368}{10} = \boxed{36.8}$.

14.18 We want

$$\frac{2+0+1+5+x}{5} \in \mathbb{Z} \iff 5 \mid x+3$$

$\therefore \min(x) = \boxed{2}$.

14.19 Since x is the mean,

$$x = \frac{60 + 100 + x + 40 + 50 + 200 + 90}{7}$$
$$= \frac{540 + x}{7}.$$

Therefore, $7x = 540 + x$, so $x = \boxed{90}$.

14.20 Let x be the sum of the integers and y be the number of elements in the list. Then we get the equations

$$\frac{x+15}{y+1} = \frac{x}{y} + 2$$

$$\frac{x+15+1}{y+1+1} = \frac{x+16}{y+2} = \frac{x}{y} + 2 - 1 = \frac{x}{y} + 1$$

. With a lot of algebra, the solution is found to be $y = \boxed{4}$.

14.21 As $a+4, 1+b, c-8$ are distinct from the three modes, each of the three modes cannot appear three times. Hence, they appear exactly twice. Our set of numbers can be organized as

$$\{a, a, b, b, c, c, a+4, b+1, c-8\}$$

Calculating the mean, we obtain

$$\frac{3(a+b+c)+(4+1-8)}{9} = 17 \implies 3(a+b+c) = \boxed{156}$$

14.22 Let $a, b (a < b)$ be the modes and x, y be the other remaining elements. Arrange them in increasing order. This yields 6 cases. Let's investigate whether each case is possible.

- $aabbxy$: Impossible. The median b is an integer, so it can't be equal to 5.5.
- $aaxbby$: Impossible. The middle two must sum to 11, so $x = 11 - b$. As the range is 5, $y = a+5$. We know the sum of all the numbers is 35, so

$$2a + 11 + b + (a+5) = 35 \iff 3a + b = 19.$$

Testing cases $(a,b) = (3,10), (4,7), (5,4), (6,1)$ we can see none of them work.

- $aaxybb$: Impossible The middle two x, y sum to 11, so $2a + 2b = 35 - 11 = 24 \implies a + b = 12$. As $b = a + 5$, $2a + 5 = 12 \implies a = \frac{7}{2}$, which makes no sense.
- $xaabby$: Possible. $a + b = 11$ so $x + y = 35 - 22 = 13$. As $y = x + 5$, this gives $x = 3$. We can observe that $3, 5, 5, 6, 6, 8$ and $3, 4, 4, 7, 7, 8$ work.
- $xaaybb$: Possible. $a + y = 11$ implies $x + a + 2b = 35 - 11 = 24$. As $x = b - 5$,

$$a + 3b = 29$$

We can observe that $b = 8, a = 5$ work as then we have $3, 5, 5, 6, 8, 8$.

- $xyaabb$: Impossible. The median a is an integer, so it can't be equal to 5.5.

Therefore, the possible values for the second largest number are $\boxed{6, 7, 8}$.

§15 Telescoping

15.1
$$\frac{1}{2} \times \frac{2}{3} \times \cdots \times \frac{19}{20} = \frac{1}{\cancel{2}} \times \frac{\cancel{2}}{\cancel{3}} \times \cdots \times \frac{\cancel{19}}{20} = \boxed{\frac{1}{20}}$$

15.2
$$\left(1+\frac{1}{1}\right)\left(1+\frac{1}{2}\right)\cdots\left(1+\frac{1}{10}\right) = \frac{2}{1} \times \frac{3}{2} \times \cdots \times \frac{11}{10}$$
$$= \frac{\cancel{2}}{1} \times \frac{\cancel{3}}{\cancel{2}} \times \cdots \times \frac{11}{\cancel{10}}$$
$$= \boxed{11}$$

15.3
$$\left(1-\frac{1}{1000}\right)\left(1-\frac{1}{999}\right)\cdots\left(1-\frac{1}{3}\right)\left(1-\frac{1}{2}\right) = \frac{999}{1000} \cdot \frac{998}{999} \cdots \frac{2}{3} \cdot \frac{1}{2}$$
$$= \frac{\cancel{999}}{1000} \cdot \frac{\cancel{998}}{\cancel{999}} \cdots \frac{\cancel{2}}{\cancel{3}} \cdot \frac{1}{\cancel{2}}$$
$$= \boxed{\frac{1}{1000}}$$

15.4
$$\left(\frac{1}{2}-\frac{\cancel{1}}{\cancel{3}}\right)+\left(\frac{\cancel{1}}{\cancel{3}}-\frac{\cancel{1}}{\cancel{4}}\right)+\cdots+\left(\frac{\cancel{1}}{\cancel{18}}-\frac{\cancel{1}}{\cancel{19}}\right)+\left(\frac{\cancel{1}}{\cancel{19}}-\frac{1}{20}\right) = \frac{1}{2}-\frac{1}{20} = \boxed{\frac{9}{20}}$$

15.5 $\frac{1}{1\times 2} + \frac{1}{2\times 3} + \cdots + \frac{1}{99\times 100}$ can be written as
$$\left(1-\frac{1}{2}\right)+\left(\frac{1}{2}-\frac{1}{3}\right)+\cdots+\left(\frac{1}{99}-\frac{1}{100}\right) = 1 - \frac{1}{100} = \boxed{\frac{99}{100}}$$

15.6
$$\frac{1}{3}+\frac{1}{15}+\frac{1}{35}+\frac{1}{63}+\frac{1}{99}+\frac{1}{143}$$
$$= \frac{1}{2}\left[\left(\frac{1}{1}-\frac{1}{3}\right)+\left(\frac{1}{3}-\frac{1}{5}\right)+\cdots+\left(\frac{1}{11}-\frac{1}{13}\right)\right]$$
$$= \frac{1}{2}\left[1-\frac{1}{13}\right] = \boxed{\frac{6}{13}}$$

15.7
$$4((-1+2)+(-3+4)+(-5+6)+\cdots+(-999+1000)) = 4(\underbrace{1+1+\cdots+1}_{500})$$
$$= 4(500)$$
$$= \boxed{2000}$$

15.8
$$(100-98)+(96-94)+(92-90)+\cdots+(8-6)+(4-2) = \underbrace{2+2+\cdots+2}_{25}$$
$$= 2(25)$$
$$= \boxed{50}$$

15.9 By Gauss Summation formulas, $A = 1 + 3 + 5 + \cdots + 2017 + 2019 = 1010^2$ and $B = 2 + 4 + 6 + \cdots + 2018 = 1009(1010)$. Therefore, our desired sum, $A - B$ is

$$1010^2 - 1009(1010) = \boxed{1010}$$

Number Theory

§16 Primes and Divisibility

16.2 Since $1 + 1 + 7 = 9$ and $9 | 9$, we can use the divisibility rule for 9. $117 \div 9 = 13$, and 13 is a prime number. Therefore, $117 = 9 \times 13 = \boxed{3 \times 3 \times 13}$.

16.3 Let $Z = \overline{abcabc}$. Then, $Z = \overline{abc} \times 1001$. Since $1001 = 7 \times 11 \times 13$, 7, 11, and 13 must be factors of Z. $\boxed{7, 11, 13}$.

16.4 Let the four-digit number be $\overline{ab28}$. Then, $\overline{ab28}$ and $\overline{82ba}$ must be divisible by 8 as 8 is a factor of 16. Using the divisibility rule for 8 on $\overline{ab28}$, $8 | \overline{b28}$ so b can be 1, 5 or 9. Since $16 | \overline{82ba}$ and $16 | 8000$, $16 | \overline{2ba}$. When $b = 1$, there is no possible a. When $b = 5, a = 6$. When $b = 9$, there is no possible a. Therefore, the four-digit number is $\boxed{6528}$.

16.5 Using the divisibility rule for 15, a five-digit flippy number should be divisible by 3 and 5. Let a five-digit flippy number be $\overline{ababa}$. Using the divisibility rule for 5, $a = 0$ or $a = 5$. However, if $a = 0$, $\overline{0b0b0}$ is not a five-digit number, so $a = 5$. Using the divisibility rule for 3, $3 | 5 + b + 5 + b + 5 = 15 + 2b$. b can be 0, 3, 6, or 9. Therefore, there are $\boxed{4}$ possible five-digit flippy numbers.

16.6 $\boxed{3195}$.

16.7 $\boxed{8}$.

16.8
$55 = 5 \times 11$
$57 = 3 \times 19$
$58 = 2 \times 29$
59 is a prime number
61 is a prime number
Therefore, $\boxed{58}$ has the smallest prime factor, which is 2.

16.9 $250 = 2 \times 125 = 2 \times 5 \times 25 = 2 \times 5 \times 5 \times 5$. Therefore, the sum of the two smallest prime factors is $2 + 5 = \boxed{7}$.

16.10 $100 = 13 \times 7 + 9$, so the smallest three digit number divisible by 13 is $13 \times 8 = 104$. Similarly, $1000 = 13 \times 76 + 12$, so the largest three digit number divisible by 13 is $13 \times 76 = 988$. Using the arithmetic sequence formula, there are $(988 - 104) \div 13 + 1 = \boxed{69}$ numbers.

16.11 The sum of two odd numbers should be an even number, but 85 is an odd number. Therefore, one of the prime number is an even number. Since 2 is the only even prime number, the two prime numbers are 2 and 83 and their product is $2 \times 83 = \boxed{166}$.

16.12 $3 | 7+4+A+5+2+B+1 = 18+A+B$ so $3 | A+B$. Also, $3 | 3+2+6+A+B+4+C = 15+A+B+C$ so $3 | A + B + C$. Since $3 | A + B, 3 | C$. Therefore, C can be $\boxed{3}$.

16.13 $9 | 2 + 0 + 1 + 8 + U = 11 + U$. U can only be 7 because $0 \leq U \leq 9$. $20187 = 8 \times 2523 + 3$. Therefore, the remainder is $\boxed{3}$.

16.14 The two smallest prime numbers greater than 50 are 53 and 59. The desired number can be obtained by multiplying them, which is $53 \times 59 = \boxed{3127}$.

16.15 $98! + 99! + 100! = 98!(1 + 99 + 99 \times 100) = 98!(10000)$. Since $10000 = 2^4 \times 5^4$, the largest n is

$\lfloor \frac{98}{5} \rfloor + \lfloor \frac{98}{25} \rfloor + 4 = \boxed{26}$. ($\because 98!$ has $\lfloor \frac{98}{5} \rfloor$ multiples of 5 and $\lfloor \frac{98}{25} \rfloor$ multiples of 25 which has additional factor of 5 that is not counted in $\lfloor \frac{98}{5} \rfloor$)

16.16 The only possible values of uniform numbers are $\{11, 13, 17\}$ because the next smallest prime number 23 exceeds 31 when added by any other uniform number. Since Caitlin's birthday is the latest, the sum of Ashley's and Brittany's uniform numbers should be $13 + 17 = 30$ and Caitlin's uniform number should be $\boxed{11}$.

16.17 $792 = 2^3 \times 3^2 \times 11$. Therefore, $\overline{AB962C}$ must be divisible by 8, 9, and 11. Using the divisibility rule for 8, $\overline{62C}$ is divisible by 8, so the only possible value for C is $C = 4$. Also, using the divisibility rule for 9, $9|A+B+9+6+2+4 = A+B+21$ and using the divisibility rule for 11, $11|A-B+9-6+2-4 = A-B+1$. Since $1 \leq A, B \leq 9$, $A + B = 6$ or 15. Similarly, $A - B = -1$ or 8. As $A + B$ and $A - B$ have same parity, there are two cases to consider.
Case 1: $A + B = 6$ and $A - B = 8$. This case cannot hold because $A = 7$ and $B = -1$.
Case 2: $A + B = 15$ and $A - B = -1$. Then $A = 7$ and $B = 8$.
Therefore, the value of A is $\boxed{7}$.

16.19 19! is divisible by 9, 11 and 16. Since there are $\lfloor \frac{19}{5} \rfloor = 3$ powers of 5 and more than three powers of 2 in 19!, 19! is a multiple of $2^3 \times 5^3 = 1000$. Hence, there are three ending zeros and $H = 0$. Using the divisibility rule for 9, $9|T + M + 33$ and using the divisibility rule for 11, $11|T - M - 7$. Therefore, $T = 4$ and $M = 8$ and the desired answer is $\boxed{12}$.

16.20 Since the number is divisible by 15, its unit digit must be 0 or 5. There are five cases to consider, depending on the number of digits.
Case 1: 1-digit number. 0 and 5 does is not multiple of 3.
Case 2: 2-digit number. Among $\{15, 25, 35, 45\}$, 15 and 45 are multiples of 3.
Case 3: 3-digit number. Among $\{125, 135, 145, 235, 245, 345\}$, 135 and 345 are multiples of 3.
Case 4: 4-digit number. Among $\{1235, 1245, 1345, 2345\}$, 1245 is a multiple of 3.
Case 5: 5-digit number. 12345 is a multiple of 3.
Therefore, the possible uphill numbers are 15, 45, 135, 345, 1245, 12345 and the answer is $\boxed{6}$.

16.21 $\boxed{420 = 2^2 \times 3^1 \times 5^1 \times 7^1}$

16.22 $\boxed{1001 = 7^1 \times 11^1 \times 13^1}$

16.23 There are 9 cases to consider, depending on the value of the unit digit.
Case 1: the unit digit is 1. 11,21,31, and 41 satisfy the property
Case 2: the unit digit is 2. 12,22,32, and 42 satisfy the property
Case 3: the unit digit is 3. 33 satisfies the property
Case 4: the unit digit is 4. 24 and 44 satisfy the property
Case 5: the unit digit is 5. 15,25,35, and 45 satisfy the property
Case 6: the unit digit is 6. 36 satisfies the property
Case 7: the unit digit is 7. No number satisfies the property
Case 8: the unit digit is 8. 48 satisfies the property
Case 9: the unit digit is 9. No number satisfies the property
Therefore, there are $4 + 4 + 1 + 2 + 4 + 1 + 1 = \boxed{17}$ numbers.

16.24 $2000 = 2 \times 1000$. The smallest multiple of 2 greater than 2000 is 2002, and 2002 is not a multiple of 3 and 5.
$2000 = 3 \times 666 + 2$. The smallest multiple of 3 greater than 2000 is 2001, and 2001 is not a multiple of 2 and 5.

$2000 = 5 \times 400$. The smallest multiple of 5 greater than 2000 is 2005, and 2005 is not a multiple of 2 and 3.
Therefore, the desired answer is $2001 + 2002 + 2005 = \boxed{6007}$

§17 Factors

17.2 $112 = 2^4 \times 7^1$. Therefore, the number of factors is $(4+1) \times (1+1) = \boxed{10}$

17.3 $144 = 2^4 \times 3^2$. A perfect square should have even powers in its prime factorizations, so let $N = 2^{2e_1} \times 3^{2e_2}$ where N is a perfect square and a factor of 144. $0 \le 2e_1 \le 4$ and $0 \le 2e_2 \le 2$ so $0 \le e_1 \le 2$ and $0 \le e_2 \le 1$. Therefore, the number of factors that are perfect squares is $(2+1) \times (1+1) = \boxed{6}$

17.4 $2020 = 2^2 \times 5^1 \times 101^1$. 2020 has $(2+1) \times (1+1) \times (1+1) = 12$ factors. Let $N = 2^{e_1} \times 5^{e_2} \times 101^{e_3}$ be a factor of 2020. We can subtract the number of factors that has less than or equal to three factors from the total number of factors.
Case 1: numbers that have one factor. $(e_1+1) \times (e_2+1) \times (e_3+1) = 1$ when $e_1 = e_2 = e_3 = 0$.
Case 2: numbers that have two factors. $(e_1+1) \times (e_2+1) \times (e_3+1) = 2$ when $(e_1, e_2, e_3) = (0,0,1), (0,1,0)$, or $(1,0,0)$.
Case 3: numbers that have three factors. $(e_1+1) \times (e_2+1) \times (e_3+1) = 3$ when $(e_1, e_2, e_3) = (2,0,0)$.
Therefore, there are $12 - 5 = \boxed{7}$ factors that have more than 3 factors.

17.6 $112 = 2^4 \times 7^1$. Therefore, the sum of factors is $(1 + 2^1 + 2^2 + 2^3 + 2^4)(1+7) = (31)(8) = \boxed{248}$.

17.8 $20 = 2^2 \times 5$. The number of factors is $(2+1) \times (1+1) = 6$. Therefore, the product of the factors is $20^{\frac{6}{2}} = \boxed{8000}$.

17.9 $54 = 2 \times 3^3$. All odd factors cannot have 2 as a factor, so they are factors of $3^3 = 27$. Hence, the product of all odd factors is $27^{\frac{(3+1)}{2}} = 729$. $36 = 2^2 \times 3^2$. The sum of all even factors is $(1 + 2 + 2^2)(1 + 3 + 3^2) - (1)(1 + 3 + 3^2) = (2 + 2^2)(1 + 3 + 3^2) = 78$ because all odd factors are factors of 9 and there sum is $(1 + 3 + 3^2)$. Therefore, $a - b = 729 - 78 = \boxed{651}$

17.10 $2010 = 2^1 \times 3^1 \times 5^1 \times 67^1$. $2 + 3 + 5 + 67 = \boxed{77}$.

17.11 $23222 = 2^6 \times 3^1 \times 11^2$. $(6+1) \times (1+1) \times (2+1) = \boxed{42}$

17.12 $98! + 99! + 100! = 98!(1 + 99 + 99 \times 100) = 98!(10000)$. Since $10000 = 2^4 \times 5^4$, the largest n is $\lfloor \frac{98}{5} \rfloor + \lfloor \frac{98}{25} \rfloor + 4 = \boxed{26}$. ($\because$ 98! has $\lfloor \frac{98}{5} \rfloor$ multiples of 5 and $\lfloor \frac{98}{25} \rfloor$ multiples of 25 which has additional factor of 5 that is not counted in $\lfloor \frac{98}{5} \rfloor$)

17.13 The question directly implies that the number has 12 factors. Let $N = p_1^{e_1} p_2^{e_2} \ldots p_k^{e_k}$ be the number. Then, $(e_1+1) \times \cdots \times (e_k+1) = 12$. There are three cases to consider depending on the value of k.
Case 1: $k = 1$. Then, $e_1 = 11$ and the smallest N is $2^{11} = 2048$.
Case 2: $k = 2$. Then, $(e_1, e_2) = (5,1)$ or $(3,2)$. The smallest N is $\min(2^5 \times 3^1, 2^3 \times 3^2) = 72$
Case 3: $k = 3$. Then $(e_1, e_2, e_3) = (2,1,1)$ The smallest N os $2^2 \times 3 \times 5 = 60$.
Therefore, the smallest number is $\boxed{60}$.

17.14 $1001 = 7^1 \times 11^1 \times 13^1$. The multiple cannot have other prime factors because $(1+1)(1+1)(1+1)(1+1) = 16 > 15$. Therefore, there are $\boxed{3}$ possible multiples: $7 \times 1001, 11 \times 1001$ and 13×1001 in which they have $(2+1)(1+1)(1+1) = 12$ factors.

17.15 $N = 2^3 \times 3^5 \times 5^1 \times 7^1 \times 17^2$. There are $(1)[(1 + 3 + \cdots + 3^5)(1+5)(1+7)(1+17+17^2)]$ odd factors and $(2+4+8)[(1+3+\cdots+3^5)(1+5)(1+7)(1+17+17^2]$ even factors because odd factors cannot have 2 as a factor. Therefore, the ratio is $\boxed{1:14}$.

17.16 $60 = 2^2 \times 3^1 \times 5^1$. $(2+1)(1+1)(1+1) = \boxed{12}$.

17.17 $320 = 2^6 \times 5$. The sum of factors is $(1 + 2^1 + \cdots + 2^6)(1+5) = (\frac{2^7-1}{2-1})(6) = \boxed{762}$ using the geometric sum formula.

§18 GCF and LCM

18.1 $180 = 2^2 \times 3^2 \times 5^1$ and $594 = 2^1 \times 3^3 \times 11^1$. The LCM is $2^2 \times 3^3 \times 5^1 \times 11 = 5940$ and the GCF is $2 \times 3^2 = 18$. Therefore, the ratio is $\frac{5940}{18} = \boxed{330}$.

18.3 $\gcd(m,n) = \frac{mn}{\text{lcm}(m,n)} = \frac{1260}{210} = \boxed{6}$.

18.5 $15 = 3 \times 5$, $22 = 2 \times 11$, and $176 = 2^4 \times 11^1$. a is the LCM of these three numbers, so $a = 2^6 \times 3^3 \times 5^3 \times 11^3$, and $b = 2^2 \times 3 \times 5 \times 11 = \boxed{660}$.

18.7 $\boxed{2}$.

18.8 $\boxed{34}$.

18.9 $\gcd(12,15) = 3$, so $b = 1$ or $b = 3$. There are two cases to consider, depending on the value of b.
Case 1: $b = 1$. Then $a = 12$ and $b = 15$, so $\text{lcm}(12,15) = 60$.
Case 2: $b = 3$. Then $a = 4$ and $b = 5$, so $\text{lcm}(4,5) = 20$.
Therefore, the answer is $\boxed{20}$.

18.10 Let the number be $\overline{a_1 a_2 \ldots a_k}$. $a_k = 2$ because the number is divisible by 2. $3 | a_1 + a_2 + \cdots + 2$ implies that there should be two more 2s. Therefore the answer is $\boxed{2232}$ as the number should contain at least one 3.

18.11 $\text{lcm}(8,9) = 72$. $10000 = 72 \times 138 + 64$, so the smallest five digit multiple of 72 is $72 \times 139 = \boxed{10008}$.

18.12 $3 - 9 + 5 - 1 = 10$, $9 - 0 + 7 - 6 + 5 - 4 = 11$, and $1 - 4 + 2 - 5 + 6 = 0$. Therefore, $\boxed{907654}$ and $\boxed{14256}$ are divisible by 11.

18.13 $10000 = 11 \times 909 + 1$. Therefore the desired answer is $11 \times 910 = \boxed{10010}$.

18.14 $1 - 2 + 3 - 4 + 5 - 6 + 7 = \boxed{4}$.

18.15 $90000 = 11 \times 8181 + 9$. Therefore, the desired answer is $11 \times 8182 = \boxed{90002}$.

§19 Modular Arithmetic

19.3 The unit digit of the desired answer is equivalent to the unit digit of $2!! + 4!! + 6!! + 8!!$ because the unit digit of $n!!$ is always 0 when $n \geq 10$. Therefore, $2 + 8 + 8 + 4 = \boxed{22}$.

19.5 $\boxed{1}$.

19.6 $\boxed{71}$.

19.7
$2^1 \equiv 2 \pmod{10}$,
$2^2 \equiv 4 \pmod{10}$
$2^3 \equiv 8 \pmod{10}$,
$2^4 \equiv 6 \pmod{10}$
$2^5 \equiv 2 \pmod{10}$
∴ We can identify the digit cycle with the length of $4 : 2, 4, 8,$ and 6. Since $1026 \equiv 2 \pmod 4$, the unit digit of 2^{1026} is $\boxed{4}$.

19.8 $13^2 \equiv 9 \equiv -1 \pmod{10}$. $13^{2012} = (13^2)^{1006} \equiv (-1)^{1006} \equiv \boxed{1} \pmod{10}$

19.9 $19 \equiv 99 \equiv -1 \pmod{10}$. $19^{19} + 99^{99} \equiv (-1)^{19} + (-1)^{99} \equiv (-1) + (-1) \equiv \boxed{8} \pmod{10}$.

19.10
$7^1 \equiv 7 \pmod{100}$
$7^2 \equiv 49 \pmod{100}$
$7^3 \equiv 43 \pmod{100}$
$7^4 \equiv 01 \pmod{100}$
$7^4 = 2401 \equiv 1 \pmod{100}$. $7^{2011} = 7^3 \times (7^4)^{502} = 43 \times 1 \equiv \boxed{43} \pmod{100}$

19.11 $\text{lcm}(3, 4, 5, 6) + 2 = \boxed{62}$.

19.12 Let $N = 9x + 1$. Then, $9x + 1 \equiv 3 \pmod{10}$. $9x \equiv 2 \implies -x \equiv 2 \implies x \equiv 8$. Therefore, $N = 9 \times 8 + 1 = 73$. $73 \equiv 7 \pmod{11}$, so the desired answer is $\boxed{7}$.

19.13 $0303 \times 0505 \equiv 3015 \pmod{10000}$. Therefore, $3 + 5 = \boxed{8}$

19.14 The question directly implies that the number has a remainder of -4 when divided by $6, 9,$ and 11. Since $\text{lcm}(6,9,11) = 198$, the possible values are $198 - 4, 2 \cdot 198 - 4, 3 \cdot 198 - 4, 4 \cdot 198 - 4, 5 \cdot 198 - 4$. Therefore, the desired answer is $\boxed{5}$.

Geometry

§20 Angle Chasing

20.2 Let x be the degree measure of angle A and y be the third angle in the triangle created by angles x and $110°$. Then
$$100 + 40 + y = 180 \implies y = 40.$$
$$x + 150 = x + y + 110 = 180 \implies x = \boxed{30}.$$

20.3 As $DE \parallel BC$, $\angle BDE = 180 - 40 = 140$. Now, as $\angle BDE = \angle BAE + \angle ABD + \angle AED$,
$$140 = \angle BAE + 10 + 80 \implies \angle BAE = \boxed{50°}$$

20.5 By angle chasing, $\angle ACF = \angle BCF - \angle BCA = 90 - \angle BCA$. As $\angle ABC = 135°$ and $BA = BC$, $\angle BCA = 22.5$. Therefore, our answer is $\angle ACF = 90 - 22.5 = \boxed{67.5}$.

20.7 As $CD = BD$, $\triangle DCB$ is isosceles. As $\angle BDC = 120 - 90 = 30$, $\angle CBD = \frac{180-30}{2} = 75$. Therefore,
$$\angle ABC = \angle ABD - \angle CBD = 120 - 75 = \boxed{45°}$$

20.14 As $AD = AB = AE$, $\triangle ADE$ is isosceles. As $\angle DAE = 90 + 60 = 150$, $\angle AED = \frac{180-150}{2} = \boxed{15°}$.

20.15 As $AD = CD$ and $\angle ADC = 60°$, $\triangle ADC$ is equilateral. Therefore, $AC = \boxed{17}$.

20.16 As $BD = DC$, $\angle DBC = \angle BCD = 70°$. Therefore, $\angle ADB = \angle DBC + \angle DCB = 70 + 70 = \boxed{140°}$.

20.17 Let X be the intersection of segment $\overline{AT}$ and circle $\odot(RSB)$. As $\angle RAS = 74°$,
$$\overset{\frown}{RB} + \overset{\frown}{SX} = 180 - 74 = 106$$
On the other hand, by **Theorem 20.9**,
$$\overset{\frown}{RB} - \overset{\frown}{SX} = 2\angle RB = 56.$$
Combining these equations give $\overset{\frown}{BR} = \frac{106+56}{2} = \boxed{81°}$.

20.18 Each divided arc makes up an angle of $30°$. Therefore,
$$\angle AOE = 4 \cdot 30 = 120°$$
$$\angle IOG = 2 \cdot 30 = 60°$$
As $\triangle AOE$ is isosceles and $\triangle IOG$ is equilateral, $x = \frac{180-120}{2} = \boxed{30}$ and $y = \boxed{60}$.

20.19 If the shared angle is x, $70 + 2x = 180 \implies x = 55$. If the shared angle is 70, $140 + x = 180 \implies x = 40$. Therefore, our answer is $\boxed{40, 55}$.

20.20 As the circles pass through each others' centers, $AE = EB = AB \implies \triangle AEB$ equilateral. Therefore, $\angle AEB = 60°$ and
$$\angle ECA = \frac{1}{2}\angle EAB = 30°, \qquad \angle EDB = \frac{1}{2}\angle EBA = 30°.$$

Therefore,
$$\angle CED = \angle AEB + \angle AEC + \angle DEB = 60 + \angle ECA + \angle EDB = \boxed{120°}.$$

20.21 The 10-sided polygon surrounded by the congruent isosceles trapezoids is made of 8 angles sized $360 - 2x$ and 2 angles sized $180 - x$. The sum of the angles in any 10-gon is $180(10 - 2) = 180 \cdot 8$. Therefore,
$$180 \cdot 8 = 8(360 - 2x) + 2(180 - x) \implies x = \boxed{100°}$$

20.22 $\angle ABD = \frac{1}{2} \cdot 70 = 35°$ and $\angle BAD = 40°$. Therefore, $\angle BDC = \angle BAD + \angle ABD = 40 + 35 = 75°$.

20.23 By Angle Chasing,
$$\angle BCA + \angle PAC + \angle PBC = \angle APB \implies \angle BCA + 20 + 30 = 90 \implies \angle BCA = \boxed{40°}$$

20.24 First compute $\angle FDE$.
$$\angle FDE = 360 - (\angle ADC + \angle EDC) = 360 - (90 + 110) = 160.$$
As $\triangle FDE$ is isosceles, $\angle DFE = 10°$. This gives $\angle AFE = 180 - 10 = \boxed{170°}$.

20.25 Since the dashed lines bisect the angles, $\angle XDA = \angle XBA = 16°$. This gives that $\angle DXB = \angle ADX + \angle ABX + \angle DAB = 84 + 16 + 16 = \boxed{116}°$.

20.26 As $\triangle ACD$ is isosceles, $\angle CDA = \frac{180-34}{2} = 73°$, which implies $\angle ADB = 180 - 73 = 107$. Now as BE bisects $\angle ABC$, $\angle DBE = \frac{1}{2}(90 - 34) = 28$. Therefore,
$$\angle DEB = 180 - (\angle EDB + \angle DBE) = 180 - (107 + 28) = \boxed{45°}.$$

20.27 As $AF = AE, AD = AB$, we have $DF = BE$ by the Pythagorean theorem on $\triangle DFA$ and $\triangle BEA$. Therefore, our equilateral triangle is symmetric with respect to diagonal CA.

As $\triangle AME$ is a $30 - 60 - 90$ triangle, $FM = MP = AP = PE = ME$, which gives $\angle QMF = \angle PME = 60$. By symmetry, $\angle BAE = \angle FAD = 15$, which implies $\angle DFA = 90 - 15 = 75$. As $\angle AFE = 60$, $\angle CFE = 180 - (\angle DFA + \angle AFE) = 180 - (75 + 60) = 45°$. Therefore, $\angle CQP = \angle QFM + \angle QMF = 60 + 45 = \boxed{105}$.

§21 Triangle

21.3 Let a, b, c be the sides of $\triangle ABC$ that correspond to heights $\frac{15}{7}, 5, 3$. Through area relations, we can get
$$\frac{15}{7}a = 5b = 3c$$
Therefore, the ratio $a : b : c = \frac{7}{15} : \frac{1}{5} : \frac{1}{3} = 7 : 3 : 5$. Let $a = 7x, b = 3x, c = 5x$ for some real x. Let $s = \frac{1}{2}(a + b + c) = \frac{15}{2}x$. By Heron's Formula,
$$[ABC]^2 = s(s-a)(s-b)(s-c) = \frac{15}{2}x \cdot \frac{1}{2}x \cdot \frac{9}{2}x \cdot \frac{5}{2}x = \frac{3^3 \cdot 5^2}{16}x^4$$
On the other hand, $[ABC] = \frac{1}{2}b \cdot 5 = \frac{15}{2}x \implies [ABC]^2 = \frac{3^2 \cdot 5^2}{4}x^2$.
$$\frac{3^3 \cdot 5^2}{16}x^4 = \frac{3^2 \cdot 5^2}{4}x^2 \implies x = \frac{2}{\sqrt{3}}$$
$$\therefore [ABC]^2 = \frac{15^2}{4} \cdot \frac{4}{3} = \boxed{75}.$$

21.15 The figure can be seen as one 3×3 box with an area 1 triangles sticking out on each side. Therefore, the answer is $\boxed{13}$.

21.16 By the Pythagorean Theorem, $AC = CB = \sqrt{1^2 + 3^2} = \sqrt{10}$. Note that the slope of AC, CB are $\frac{1}{3}, -3$, respectively, which multiply to (-1). This implies $\angle ACB = 90°$. Therefore, $[ABC] = \frac{1}{2} \cdot \sqrt{10} \cdot \sqrt{10} = 5$. Since the total area of the grid is 30, this gives that the triangle is $\boxed{\dfrac{1}{6}}$ the area of the grid.

21.17
$$\frac{\frac{1}{4} + \frac{1}{8} + \frac{3}{8} + \frac{1}{4}}{4} = \boxed{\frac{1}{4}}$$

21.18 Because $\overline{BD} + \overline{CD} = 5$, we can see that when we draw a line from point B to imaginary point D that line applies to both triangles. Let us say that x is that line. Perimeter of $\triangle ABD$ would be $\overline{AD} + 4 + x$, while the perimeter of $\triangle ACD$ would be $\overline{AD} + 3 + (5 - x)$. Notice that we can find x from these two equations by setting them equal and then canceling $\overline{AD}$. We find that $x = 2$, and because the height of the triangles is the same, the ratio of the areas is $2 : 3$, so that means that the area of $\triangle ABD = \frac{2 \cdot 6}{5} = \boxed{\dfrac{12}{5}}$.

21.19 We first connect point B with point D.

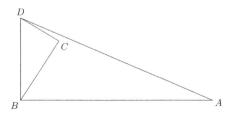

We can see that $\triangle BCD$ is a 3-4-5 right triangle. We can also see that $\triangle BDA$ is a right triangle, by the 5-12-13 Pythagorean triple. With these lengths, we can solve the problem. The area of $\triangle BDA$ is $\frac{5 \cdot 12}{2}$, and the area of $\triangle BCD$ is $\frac{3 \cdot 4}{2}$. Thus, the area of quadrilateral $ABCD$ is $30 - 6 = \boxed{24}$. **21.20** By area formula,

$$[ADC] = \frac{1}{2} \cdot 3 \cdot 3 = \boxed{\frac{9}{2}}$$

21.21 The area of $\triangle BFD$ is the area of square $ABCE$ subtracted by the the area of the three triangles around it. Arbitrarily assign the side length of the square to be 6.

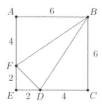

The ratio of the area of $\triangle BFD$ to the area of $ABCE$ is

$$\frac{36 - 12 - 12 - 2}{36} = \frac{10}{36} = \boxed{\frac{5}{18}}$$

21.22 The area of $\triangle CDE$ is $\frac{DC \cdot CE}{2}$. The area of $ABCD$ is $AB \cdot AD = 5 \cdot 6 = 30$, which also must be equal to the area of $\triangle CDE$, which, since $DC = 5$, must in turn equal $\frac{5 \cdot CE}{2}$. Through transitivity, then, $\frac{5 \cdot CE}{2} = 30$, and $CE = 12$. Then, using the Pythagorean Theorem, you should be able to figure out that $\triangle CDE$ is a $5 - 12 - 13$ triangle, so $DE = \boxed{13}$.

21.23 The area of the square around the pinwheel is 25. The area of the pinwheel is equal to the square − the white space. Each of the four triangles have a base of 3 units and a height of 2.5 units, and so their combined area is 15 units squared. Then the unshaded space consists of the four triangles with total area of 15, and there are four white corner squares. Therefore the area of the pinwheel is $25 - (15 + 4)$ which is $\boxed{6}$.

21.24 Graphing the lines we can see that the height of the triangle is 4, and the base is 8. Using the formula for the area of a triangle, we get $\frac{4 \cdot 8}{2}$ which is equal to $\boxed{16}$.

21.25 Draw X on $\overline{AF}$ such that $\overline{XD}$ is parallel to $\overline{BC}$. Triangles BEF and EXD are similar, and since $BE = ED$, they are also congruent, and so $XE = EF$ and $XD = BF$. $AD : AC = 3$ implies $\frac{AX}{AF} = 3 = \frac{FC}{XD} = \frac{FC}{BF}$, so $BC = BF + 3BF = 4BF$, $BF = \frac{BC}{4}$. Since $XE = EF$, $AX = XE = EF$, and since $AX + XE + EF = AF$, all of these are equal to $\frac{AF}{3}$, and so the altitude of triangle BEF is equal to $\frac{1}{3}$ of the altitude of ABC. The area of ABC is 360, so the area of $\triangle EBF = \frac{1}{3} \cdot \frac{1}{4} \cdot 360 = \boxed{30}$.

§22 Quadrilateral

22.1 The area of the shaded region is $(24s)^2$. To find the area of the large square, we note that there is a d-inch border between each of the 23 pairs of consecutive squares, as well as from between first/last squares and the large square, for a total of $23 + 2 = 25$ times the length of the border, i.e. $25d$. Adding this to the total length of the consecutive squares, which is $24s$, the side length of the large square is $(24s + 25d)$, yielding the equation $\frac{(24s)^2}{(24s+25d)^2} = \frac{64}{100}$. Taking the square root of both sides (and using the fact that lengths are non-negative) gives $\frac{24s}{24s+25d} = \frac{8}{10} = \frac{4}{5}$, and cross-multiplying now gives $120s = 96s + 100d \implies 24s = 100d \implies \frac{d}{s} = \frac{24}{100} = \boxed{\frac{6}{25}}$.

22.2 As $15^2 + 20^2 = 25^2$ and $20^2 + 21^2 = 29^2$, we can see that $\triangle RAE$ and $\triangle PEA$ are right triangles. Therefore, the area of the quadrilateral is

$$\frac{1}{2} \cdot 15 \cdot 20 + \frac{1}{2} \cdot 20 \cdot 21 = \boxed{360}.$$

22.3 We can see that there are 2 rectangles lying on top of the other and that is the same as the length of one rectangle. Now we know that the shorter side is 5, if we take the information of the problem, and the bigger side is 10, if we do $5 \cdot 2 = 10$. Now we get the sides of the big rectangles being 15 and 10, so the area is $\boxed{150}$.

22.4 Drawing segments AC and BD, the number of triangles outside square $ABCD$ is the same as the number of triangles inside the square. Thus areas must be equal so the area of $ABCD$ is half the area of the larger square which is $\frac{60}{2} = \boxed{30}$.

22.5 If you draw altitudes from A and B to CD, the trapezoid will be divided into two right triangles and a rectangle. Let X, Y be the foot of perpendicular from A, B to $\overline{CD}$. You can find the values of a and b with the Pythagorean theorem.

$$a = \sqrt{15^2 - 12^2} = \sqrt{81} = 9$$

$$b = \sqrt{20^2 - 12^2} = \sqrt{256} = 16$$

$ABYX$ is a rectangle so $XY = AB = 50$.

$$CD = a + XY + b = 9 + 50 + 16 = 75$$

The area of the trapezoid is

$$12 \cdot \frac{(50 + 75)}{2} = 6(125) = \boxed{750}$$

22.6 The midpoints of the four sides of every rectangle are the vertices of a rhombus whose area is half the area of the rectangle: Note that the diagonals of the rhombus have the same lengths as the sides of the rectangle.

Let $A = (-3, 0), B = (2, 0), C = (5, 4),$ and $D = (0, 4)$. Note that $A, B, C,$ and D are the vertices of a rhombus whose diagonals have lengths $AC = 4\sqrt{5}$ and $BD = 2\sqrt{5}$. It follows that the dimensions of the rectangle are $4\sqrt{5}$ and $2\sqrt{5}$, so the area of the rectangle is $4\sqrt{5} \cdot 2\sqrt{5} = \boxed{40}$.

§23 Circles

23.1 As $AB = 1$, $AX = \frac{1}{2}$ and $WX = \frac{\sqrt{2}}{2}$. Therefore, each circle has a radius of $\frac{\sqrt{2}}{4}$. So, each semicircle has an area of $\pi \cdot \left(\frac{\sqrt{2}}{4}\right)^2 = \frac{\pi}{8}$. Hence, the area of region R is $1 - \frac{1}{2}\pi$, giving an answer of $\boxed{\frac{3}{2}}$.

23.2 Observe $\triangle OQN$. As $OQ = 4 = 2 \cdot ON$, $\angle QON = 60°$. Hence, we have a $30 - 60 - 90$ triangle, so $QN = 2\sqrt{3}$ and $QR = 2(\sqrt{3} - 1)$. Now we're ready to calculate our desired area. We'll subtract the areas outside QRP from sector AOB.

Sector AOB has area $\pi \cdot 4^2 \cdot \frac{1}{4} = 4\pi$. The area of sector AOQ and POB are each $\pi \cdot 4^2 \cdot \frac{1}{12} = \frac{4}{3}\pi$ as their angle is $30°$. The area of triangles OQR and ORP are each $\frac{1}{2} \cdot 2 \cdot (2\sqrt{3} - 2) = 2\sqrt{3} - 2$. Therefore, our desired area is

$$4\pi - \left(2 \cdot \frac{4}{3}\pi + 2(\sqrt{3} - 2)\right) = \boxed{\frac{4}{3}\pi - 4\sqrt{3} + 4}$$

23.3 Let O be the center of the semicircle. The diameter of the semicircle is $9 + 16 + 9 = 34$, so $OC = 17$. By symmetry, O is in fact the midpoint of DA, so $OD = OA = \frac{16}{2} = 8$. By the Pythagorean theorem in right-angled triangle ODC (or OBA), we have that CD (or AB) is $\sqrt{17^2 - 8^2} = 15$. Accordingly, the area of $ABCD$ is $16 \cdot 15 = \boxed{240}$.

23.4 Draw the common perpendicular bisector of the two chords and suppose it intersects the longer chord at A and the shorter chord at B. Let X be one of the points where the longer chord intersects the circle, and let Y be one of the points where the shorter chord intersects the circle.

Then by Pythagorean Theorem on $\triangle CAX$, $CA = \sqrt{25^2 - 24^2} = 7$. Let x be the distance between the two chords. Using Pythagorean Theorem again on $\triangle CBY$,

$$(7 + x)^2 + 12^2 = 25^2 \implies x = \sqrt{481} - 7 \approx \boxed{15}$$

23.5 Let the radius of the large circle be R. Then, the radii of the smaller circles are $\frac{R}{2}$. The areas of the circles are directly proportional to the square of the radii, so the ratio of the area of the small circle to the large one is $\frac{1}{4}$. This means the combined area of the 2 smaller circles is half of the larger circle, and therefore the shaded region is equal to the combined area of the 2 smaller circles, which is $\boxed{1}$.

23.6 Since $\triangle ACD$ is isosceles, CB bisects AD. Thus $AB = BD = 8$. From the Pythagorean Theorem, $CB = 6$. Thus the area between the two circles is $100\pi - 36\pi = 64\pi$ $\boxed{64\pi}$

23.7 The area of the smaller square is one half of the product of its diagonals. Note that the distance from a corner of the smaller square to the center is equivalent to the circle's radius so the diagonal is equal to the diameter: $2 \cdot 2 \cdot \frac{1}{2} = 2$.

The circle's shaded area is the area of the smaller square subtracted from the area of the circle: $\pi - 2$.

If you draw the diagonals of the smaller square, you will see that the larger square is split 4 congruent half-shaded squares. The area between the squares is equal to the area of the smaller square: 2.

Approximating π to 3.14, the ratio of the circle's shaded area to the area between the two squares is about

$$\frac{\pi - 2}{2} \approx \frac{3.14 - 2}{2} = \frac{1.14}{2} \approx \boxed{\frac{1}{2}}$$

23.8 The area in the rectangle but outside the circles is the area of the rectangle minus the area of all three of the quarter circles in the rectangle.

The area of the rectangle is $3 \cdot 5 = 15$. The area of all 3 quarter circles is $\frac{\pi}{4} + \frac{\pi(2)^2}{4} + \frac{\pi(3)^2}{4} = \frac{14\pi}{4} = \frac{7\pi}{2}$. Therefore the area in the rectangle but outside the circles is $15 - \frac{7\pi}{2}$. π is approximately $\frac{22}{7}$, and substituting that in will give $15 - 11 = \boxed{4.0}$.

23.9 First, we notice half a square so first let's create a square. Once we have a square, we will have a full circle. This circle has a diameter of 4 which will be the side of the square. The area would be $4 \cdot 4 = 16$. Divide 16 by 2 to get the original shape and you get $\boxed{8}$.

23.10 The entire circle's area is 144π. The area of the black regions is $(100-64)\pi + (36-16)\pi + 4\pi = 60\pi$. The percentage of the design that is black is $\frac{60\pi}{144\pi} = \frac{5}{12} \approx \boxed{42}$.

23.11 By the Pythagorean Theorem, the radius of the larger circle turns out to be $\sqrt{1^2 + 1^2} = \sqrt{2}$. Therefore, the area of the larger circle is $(\sqrt{2})^2\pi = 2\pi$. Using the coordinate plane given, we find that the radius of each of the two semicircles to be 1. So, the area of the two semicircles is $1^2\pi = \pi$. Finally, the ratio of the combined areas of the two semicircles to the area of circle O is $\boxed{\dfrac{1}{2}}$.

23.12

Draw a square around the star figure. The side length of this square is 4, because the side length is the diameter of the circle. The square forms 4-quarter circles around the star figure. This is the equivalent of one large circle with radius 2, meaning that the total area of the quarter circles is 4π. The area of the square is 16. Thus, the area of the star figure is $16 - 4\pi$. The area of the circle is 4π. Taking the ratio of the two areas, we find the answer is $\boxed{\dfrac{4-\pi}{\pi}}$.

23.13 Let the center of the semicircle be O. Let the point of tangency between line AB and the semicircle be F. Angle BAC is common to triangles ABC and AFO. By tangent properties, angle AFO must be 90 degrees. Since both triangles ABC and AFO are right and share an angle, AFO is similar to ABC. The hypotenuse of AFO is $12 - r$, where r is the radius of the circle. The short leg of AFO is r. Because $\triangle AFO \sim \triangle ABC$, we have $r/(12-r) = 5/13$ and solving gives $r = \boxed{\dfrac{10}{3}}$.

23.14 We have $\triangle ABC$ such that $AB = AC = 17$. Let D be the midpoint of BC. Draw a perpendicular from point D to AC and call that point E. We also know due to tangents that this is is r. Thus we have $\triangle DCA \sim \triangle ECD$, so

$$\frac{15}{17} = \frac{r}{8} \implies r = \boxed{\dfrac{120}{17}}$$

23.15 Let the centers of the circles containing arcs $\overset{\frown}{SR}$ and $\overset{\frown}{TR}$ be X and Y, respectively. Extend $\overline{US}$ and $\overline{UT}$ to X and Y, and connect point X with point Y.

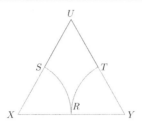

We can clearly see that $\triangle UXY$ is an equilateral triangle, because the problem states that $m\angle TUS = 60°$. We can figure out that $m\angle SXR = 60°$ and $m\angle TYR = 60°$ because they are $\frac{1}{6}$ of a circle. The area of the figure is equal to $[\triangle UXY]$ minus the combined area of the 2 sectors of the circles (in red). Using the area formula for an equilateral triangle, $\frac{a^2\sqrt{3}}{4}$, where a is the side length of the equilateral triangle, $[\triangle UXY]$ is $\frac{\sqrt{3}}{4} \cdot 4^2 = 4\sqrt{3}$. The combined area of the 2 sectors is $2 \cdot \frac{1}{6} \cdot \pi r^2$, which is $\frac{1}{3}\pi \cdot 2^2 = \frac{4\pi}{3}$. Thus, our final answer is $\boxed{4\sqrt{3} - \dfrac{4\pi}{3}}$.

23.16 Let O be the center of the circle, and let A_1 and B_1 the midpoints of the chords with length 38 (The one on the same side of O as the chord with length 34) and 34, respectively. Additionally, let A_2 and B_2 be the intersections of the chords with length 38 (The same one as the one with midpoint A_1) and 34 with the circle, respectively. Furthermore, let $2d$ be the common distance between the three chords, and let r be the radius of this circle.

Then $OA_1 = d$, $OB_1 = 3d$, and $OA_2 = OB_2 = r$. Also, $AA_1 = 19$, $BB_1 = 17$, and $\angle OA_1A_2 = \angle OB_1B_2 = 90°$; applying the Pythagorean Theorem on $\triangle OA_1A_2$ and $\triangle OB_1B_2$ gives

$$\begin{cases} d^2 + 19^2 = d^2 + 361 = r^2, \\ 9d^2 + 17^2 = 9d^2 + 289 = r^2. \end{cases}$$

Subtracting the first equation from the second, we have

$$8d^2 - 72 = 0 \implies 8d^2 = 72 \implies d^2 = 9 \implies d = 3.$$

Then, the distance between two adjacent parallel lines is $2d$, which is $\boxed{6}$.

23.17 A diameter of the circle is equal in length to the diagonal of the rectangle. By Pythagorean Theorem,
$$(2r)^2 = 10^2 + 24^2 \implies 2r = 26.$$
Therefore, the circumference, is $2\pi r = \boxed{26\pi}$.

§24 Similarity and Congruence

24.1 Let G be on $\overrightarrow{AD}$ such that $\triangle BGA$ is congruent to $\triangle BFC$. As $\angle ABE + \angle FBC = 45°$,
$$\angle GBE = \angle GBA + \angle ABE = \angle BFC + \angle ABE = 45°$$
Now as $GB = BG$ and EB is shared, $\triangle GBE \cong \triangle FBE$ (SAS). Therefore,
$$\overline{EF} = \overline{GE} = \overline{GA} + \overline{AE} = \overline{CF} + \overline{AE}$$
Let $AB = x$. Then $AB = x - 12$ and $CF = x - 35$. By the Pythagorean Theorem,
$$12^2 + 35^2 = \overline{EF}^2 = (2x - 47)^2 \implies x = \boxed{42}.$$

24.3 As $CF \parallel ME$, $\triangle BME \sim \triangle BCF$ with a ratio of $1:2$. Similarly, as $DF \parallel ME$, $\triangle ADE \sim \triangle AME$ with ratio $1:2$. Therefore, $AF = FE = EB$, giving an answer of $\boxed{\dfrac{11}{3}}$.

24.4 As $EF \parallel BC$, $\triangle AEF \sim \triangle ACB$. Based on the ratios of the side lengths, write lengths $AE = 13a, EF = 5a, AF = 12a$ for some positive real a. Now as the perimeter of $\triangle AEF$ and $\square EFBC$ are same,
$$5 + (13 - 13a) + (12 - 12a) = BC + CE + BF = AE + AF = 13a + 12a \implies a = \frac{3}{5}$$
Therefore, $EF = 5a = \boxed{3}$.

24.5 We see that half the area of the octagon is 5. We see that the triangle area is $5 - 1 = 4$. That means that $\frac{5h}{2} = 4 \to h = \frac{8}{5}$.
$$\overline{QY} = \frac{8}{5} - 1 = \frac{3}{5}$$
Therefore, our desired answer is
$$\frac{\frac{2}{5}}{\frac{3}{5}} = \boxed{\frac{2}{3}}$$

24.6 The area of trapezoid $CBFE$ is $\frac{1+3}{2} \cdot 4 = 8$. Next, we find the height of each triangle to calculate their area. The triangles are similar, and are in a $3:1$ ratio by AA similarity (alternate interior and vertical angles) so the height of the largere is 3, while the height of the smaller one is 1. Thus, their areas are $\frac{1}{2}$ and $\frac{9}{2}$. Subtracting these areas from the trapezoid, we get $8 - \frac{1}{2} - \frac{9}{2} = \boxed{3}$.

24.7 We draw a square as shown:

We wish to find the area of the biggest square. The area of this square is composed of the center white square and the four red triangles. Because the inner square has an area of $(5-2)$, 3, squared, 9, it also has a length of $\sqrt{9} = 3$. The heights of each of the red triangles are 1 (because the gray squares have lengths of one), and the area of one triangle is namely $\frac{3 \cdot 1}{2}$. Thus, the combined area of the four triangles is $4 \cdot \frac{3}{2} = 6$. Furthermore, the area of the smaller square is 9. We add these to see that the area of the large square is $9 + 6 = \boxed{15}$.

§25 Area of Polygon

25.3 Since octagon $ABCDEFGH$ is a regular octagon, it is split into 8 equal parts, such as triangles $\triangle ABO, \triangle BCO, \triangle CDO$, etc. These parts, since they are all equal, are $\frac{1}{8}$ of the octagon each. The shaded region consists of 3 of these equal parts plus half of another, so the fraction of the octagon that is shaded is $\frac{1}{8} + \frac{1}{8} + \frac{1}{8} + \frac{1}{16} = \boxed{\frac{7}{16}}$.

25.4 The area outside the small triangle but inside the large triangle is $16 - 1 = 15$. This is equally distributed between the three trapezoids. Each trapezoid has an area of $15/3 = \boxed{5}$.

25.5 If the two rectangles were seperate, the perimeter would be $2(2(2+4)) = 24$. It easy to see that their connection erases 2 from each of the rectangles, so the final perimeter is $24 - 2 \times 2 = \boxed{20}$.

25.6 The six equilateral triangular extensions fit perfectly into the hexagon meaning the answer is $\boxed{1:1}$

25.7 Notice that $AF + DE = BC$, so $DE = 4$. Let O be the intersection of the extensions of AF and DC, which makes rectangle $ABCO$. The area of the polygon is the area of $FEDO$ subtracted from the area of $ABCO$.
$$\text{Area} = 52 = 8 \cdot 9 - EF \cdot 4$$
Solving for the unknown, $EF = 5$, therefore $DE + EF = 4 + 5 = \boxed{9}$.

25.8 We can set a proportion:
$$\frac{AD}{AB} = \frac{3}{2}$$
We substitute AB with 30 and solve for AD.
$$\frac{AD}{30} = \frac{3}{2}$$
$$AD = 45$$
We calculate the combined area of semicircle by putting together semicircle AB and CD to get a circle with radius 15. Thus, the area is 225π. The area of the rectangle is $30 \cdot 45 = 1350$. We calculate the ratio:
$$\frac{1350}{225\pi} = \frac{6}{\pi} \Rightarrow \boxed{6\pi}$$

25.9 Since triangle ABD is congruent to triangle ECD and $\overline{CE} = 11$, $\overline{AB} = 11$. Since $\overline{AB} = \overline{BC}$, $\overline{BC} = 11$. Because point D is the midpoint of $\overline{BC}$, $\overline{BD} = \frac{\overline{BC}}{2} = \frac{11}{2} = \boxed{5.5}$.

25.10 Draw altitudes from B and C to base AD to create a rectangle and two right triangles. The side opposite BC is equal to 50. The bases of the right triangles can be found using Pythagorean or special triangles to be 18 and 7. Add it together to get $AD = 18 + 50 + 7 = 75$. The perimeter is $75 + 30 + 50 + 25 = \boxed{180}$.

Essential Academy (June 2023) Introduction to AMC 8: Solutions Manual

§26 3D Geometry

26.3 While imagining the folding, $\overline{AB}$ goes on $\overline{BC}$, $\overline{AH}$ goes on $\overline{CI}$, and $\overline{EF}$ goes on $\overline{FG}$. So, $BJ = CI = 8$ and $FG = BC = 8$. Also, $\overline{HJ}$ becomes an edge parallel to $\overline{FG}$, so that means $HJ = 8$.

Since $GH = 14$, then $JG = 14 - 8 = 6$. So, the area of $\triangle BJG$ is $\frac{8 \cdot 6}{2} = 24$. If we let $\triangle BJG$ be the base, then the height is $FG = 8$. So, the volume is $24 \cdot 8 = \boxed{192}$.

26.10 Using the formula for the volume of a cylinder, we get that the volume of Alex's can is $3^2 \cdot 12 \cdot \pi$, and that the volume of Felicia's can is $6^2 \cdot 6 \cdot \pi$. Now, we divide the volume of Alex's can by the volume of Felicia's can, so we get $\frac{1}{2}$, which is $\boxed{1:2}$.

26.11 There are $10 \cdot 12 = 120$ cubes on the base of the box. Then, for each of the 4 layers above the bottom (as since each cube is 1 foot by 1 foot by 1 foot and the box is 5 feet tall, there are 4 feet left), there are $9 + 11 + 9 + 11 = 40$ cubes. Hence, the answer is $120 + 4 \cdot 40 = \boxed{280}$.

26.12 The slice is cutting the cylinder into two equal wedges with equal area. The cylinder's volume is $\pi r^2 h = \pi (4^2)(6) = 96\pi$. The area of the wedge is half this which is $48\pi \approx \boxed{151}$.

26.13 The total area of the four congruent triangles formed by the squares is $5 - 4 = 1$. Therefore, the area of one of these triangles is $\frac{1}{4}$. The height of one of these triangles is a and the base is b. Using the formula for area of the triangle, we have $\frac{ab}{2} = \frac{1}{4}$. Multiply by 2 on both sides to find that the value of ab is $\boxed{\dfrac{1}{2}}$.

26.14 Note that $EJCI$ is a rhombus by symmetry. Let the side length of the cube be s. By the Pythagorean theorem, $EC = s\sqrt{3}$ and $JI = s\sqrt{2}$. Since the area of a rhombus is half the product of its diagonals, the area of the cross section is $\frac{s^2\sqrt{6}}{2}$. This gives $R = \frac{\sqrt{6}}{2}$. Thus, $R^2 = \boxed{\dfrac{3}{2}}$.

26.15 If the sea level rises 4 meters, the remains of Turkey will be a cone of height 8 and base radius of 6. This is due to similar triangles as $8 : 6 = 12 : 9$. Therefore, we need to find the surface area of this cone minus the base area.

The surrounding surface area of the cone can be unfolded into a sector of a circle. This sector has radius $\sqrt{6^2 + 8^2} = 10$ and arc length $2\pi \cdot 6 = 12\pi$. Therefore, the area of this sector is

$$\pi 10^2 \cdot \frac{12\pi}{2\pi \cdot 10} = \boxed{60\pi}.$$

26.16 The cylinder can be "unwrapped" into a rectangle, and we see that the stripe is a parallelogram with base 3 and height 80. Thus, we get $3 \times 80 = 240 \implies \boxed{240}$.

Made in the USA
Las Vegas, NV
25 September 2024

95764844R00125